FLORILEGIUM COLUMBIANUM

Photo by Joe Pineiro

Paul Oskar Kristeller

FLORILEGIUM COLUMBIANUM

ESSAYS IN HONOR OF
PAUL OSKAR KRISTELLER

EDITED BY
KARL-LUDWIG SELIG
AND
ROBERT SOMERVILLE

ITALICA PRESS
NEW YORK
1987

Paperback and Ebook Editions © 2005 by Italica Press
Italica Press, Inc.
595 Main Street
New York, NY 10044

Library of Congress Cataloging-in-Publication Data

Florilegium Columbianum.

Includes bibliographical references and index.
1. Humanism. 2. Civilization, Ancient. 3. Literature, Ancient—History and criticism. 3. Literature, Medieval—History and criticism. 4. Renaissance. 5. Kristeller, Paul Oskar, 1905- . I. Kristeller, Paul Oskar, 1905- . II. Selig, Karl-Ludwig. III. Somerville, Robert, 1940- .
CB311.F58 1987 001.3 87-80192

ISBN-13: 978-0-934977-48-7 (Paperback)

ISBN-10: 1-59910-011-8 (Ebook)
ISBN-13: 978-1-59910-011-1 (Ebook)

Printing 5 4 3 2 1

Visit:
www.ItalicaPress.com

This volume, dedicated to Paul Oskar Kristeller, is a collection of essays by senior colleagues at Columbia University who have not had the privilege of participating in earlier volumes honoring this eminent scholar.

CONTENTS

■

FOREWORD

THE INTELLECTUAL MIGRATION precipitated by the reign of Nazi terror in Europe in the 1930s and 1940s brought to this country an extraordinarily gifted generation of émigré scholars. Among those, Columbia University was privileged, in the fall of 1939, to attract to its graduate faculties a young humanist scholar and philosopher of the Renaissance, Paul Oskar Kristeller æ a man whose breadth of intellectual curiosity and knowledge, then as now, transcends the traditional disciplinary boundaries of philosophy, history, classics, art, and literature.

A measure of Kristeller's extraordinary claim to distinction may be gleaned from the fact that in an era in which the tradition of a publication of a celebratory Festschrift has been waning, he should be honored with a series of such tributes. It is particularly fitting that Columbia University, which became his intellectual home for half a century, should choose to pay one more tribute to his erudition and intellectual leadership as a scholar par excellence of Western humanist thought. As a member of Columbia University's Graduate Faculties of Arts and Sciences, Paul Kristeller has established new bench marks for standards of scholarship, research productivity, commitment to teaching, and responsive citizenship. His career provides a brilliant illustration of the teacher-scholar; one whose path-breaking research productivity has been integral to his success as a teacher and mentor of more than one generation of graduate students. His talent for organizing collaborative research projects has been attested to in a series of superbly edited texts of previously unpublished or rare manuscripts in major European libraries. His genius as a faculty colleague has found ample expression in his team teaching with such former colleagues and longtime friends as John H. Randall, Jr. and Ernest A. Moody, in fields that range from Plato to Kant, to research techniques in late medieval and Renaissance philosophy.

While Kristeller's leadership in such professional associations as the Renaissance Society of America, the Medieval Academy of America, and the Société Internationale pour l'Étude de la Philosophie Médiévale served to enhance still further his stature in the international community of medieval and Renaissance scholarship, these activities never interfered with his ongoing commitment to teaching and to editing — often in extraordinary detail — the papers and manuscripts of Columbia students colleagues, and friends.

In an age in which the very concept of humanism has become in certain circles politically and religiously suspect, Kristeller's nuanced delineation of Renaissance concepts of man serves to remind us of the critical importance of the work of historians of ideas and of philosophical thought in transmitting to future generations a fuller, more complex, and more balanced expression of these ideas and of the cultural context of which they form a part.

Gillian Lindt, Dean
Graduate School of Arts and Sciences
Columbia University
March 1987

I
Cicero's Attitude Towards Stoicism and Skepticism in the *De natura deorum*

LEONARDO TARÁN

CICERO'S RELIGIOUS CONCEPTIONS[1] — including his attitude toward the Roman state religion, whether or not a supreme being (or beings) controls and guides the universe, and the question of the immortality of the soul — have received divergent interpretations at least since the days of the Latin Christian apologists.[2] Some have taken him to be sincere in his views on theological matters; others have maintained that for reasons of political expediency and/or fear he in fact tried to conceal his unbelief; and still others have accused him of inconsistency or worse. It is not the purpose of this paper to examine the whole question of Cicero's religious beliefs but rather to discuss his opinions as expressed in the *De natura deorum* and his purpose in composing that work. I trust, however, that if my views are basically correct, they will help to clarify the whole issue. For Cicero's statement at the end of the *De natura deorum* — that he found the Stoic Balbus' argument more probable than

1. The following works will be cited by author's name: J.B. Mayor, *M. Tullii Ciceronis De Natura Deorum Libri Tres,* with introduction and commentary, 3 vols. (Cambridge, 1880, 1883, 1885); A.S. Pease, *M. Tulli Ciceronis De Natura Deorum Libri III,* 2 vols. (Cambridge, MA, 1955, 1958); J.S. Reid, *M. Tulli Ciceronis, Academica* (London, 1885); R. Philippson, "M. Tullius Cicero (29)...Die Philosophische Schriften," in *Paulys Realencyclopädie der Classischen Altertumswissenshaft* 7 A 1 (1939), cols. 1104–92. In addition, A.S. Pease, "The Conclusion of Cicero's *De Natura Deorum,*" *Transactions of the American Philological Association* 44 (1913): 25–37 will be cited as Pease (1913). Since in this paper I have often disagreed with several of Pease's interpretations, I would like here to acknowledge my indebtedness to the erudite learning of his publications on Cicero's philosophical works.
2. In addition to the references given later in this paper, see Pease 1:9 ff.; P. Defourny, "Les fondements de la religion d'après Cicéron," *Les Études Classiques* 22 (1954): 241–53, and 366–78; J.-M. André, "La philosophie religieuse de Cicéron. Dualisme Académique et Tripartition Varronienne," *Ciceroniana: Hommages à K. Kumaniecki* (Leiden, 1975), 11–21.

that of the Academic Cotta[3] — has played an important role in the controversy concerning his religious views. It has also given rise to the same kind of divergent interpretations that his religious views in general have received. Before coming to the immediate object of this paper, it will be necessary to place before the reader a few facts and theories relevant to the subsequent discussion.

CICERO CONCEIVED the plan of writing a series of philosophical works that would cover the whole field of philosophy. This happened at the latest between his forced withdrawal from public life in 46 and the death of his daughter Tullia in February 45 B.C.E.,[4] though it may have been earlier. Following his own account in *De diuinatione* 2.1-3, we know that to this corpus belong (among others) the *Consolatio*, the *Hortensius*, the *Academica*, the *De finibus,* and the *Tusculanae disputationes*, all written in 45 B.C.E. and before the *De natura deorum.* This one came next and was itself followed by the *De diuinatione*, published after the death of Caesar in March 44 B.C.E.[5] In this passage of *De diuinatione* Cicero announces his plan to write a third "theological" work, the *De fato,* which he later did write, though the work is now extant in fragmentary form.[6] We may surmise, then, that the *De natura deorum* was written in the latter part of 45, though it may have been finished in early 44 B.C.E.[7]

Like the rest of Cicero's philosophical works, the *De natura deorum* is a dialogue. It is in three books and the narrator is Cicero. Book 1 begins with a proem by Cicero himself (1.1-17). There, besides the dedication to Brutus, he emphasizes the importance of the question of the gods and defends the Academic school of

3. Cf. 3.95. (The *De natura deorum* is simply referred to by book and chapter numbers.) This statement is discussed below.
4. Cf. 1.7–9; Reid, 1–10, 20–28, 47; S.S. Häfner, "Die literarischen Pläne Ciceros," Dissertation (Munich, Coburg, 1928), esp. 94–103; O. Plasberg, *Cicero in seinen Werken und Briefen* (Leipzig, 1926), 157–69; Philippson, cols. 1123–73; Pease, 1:1–9; H.A.K. Hunt, *The Humanism of Cicero* (Melbourne, 1954), 1–15; P. Boyancé, *Études sur l'humanisme cicéronien* (1936; Brussels, 1970), 200–201.
5. Cf. *De diuinatione* 1.26–27; 1.119; 2.7; 2.23; 2.99; 2.110; etc., which presuppose that the work was published after the Ides of March 44. Of course there are many passages clearly written before Caesar's murder, but there is no need here to enter into the complicated question of the date of the different passages of the *De diuinatione*, on which cf. R. Durand, "La date du *De Diuinatione*," in *Mélanges Boissier* (Paris, 1903), 173–83; and A.S. Pease, *M. Tulli Ciceronis De Diuinatione Libri Duo,* University of Illinois Studies in Language and Literature 6 (Urbana, IL, 1920), 171–73.
6. Cf. A. Yon, *Cicéron. Traité du destin* (Paris, 1950), ii–v and xvi–xvii.
7. We have no definite evidence, except the statement in *De diuinatione* 2.1–3.

philosophy to which he belongs. He also tells how he came to the house of another Academic, Cotta, during the *feriae Latinae,* and found his host engaged in a discussion on the nature of the gods with the Epicurean Velleius and the Stoic Balbus.[8] Since Cicero had just arrived, Velleius recommences his exposition of Epicurean theology. His speech comprises an attack against Plato and the Stoa (1.18-24), followed by a doxographical account and criticism of the views of Greek philosophers from Thales to Diogenes of Babylon (1.25-41), which ends with a brief attack on the poets and on popular and Oriental superstition (1.41-43). The last part of the speech (1.43-56) is devoted to a straightforward exposition of Epicurean theology, which is considerably longer (1.57-124) than Velleius' short account. In Book 2, after a few transitional paragraphs (2.1-3), we find Balbus' long exposition of Stoic theology (2.4-168), while in Book 3, again after a few transitional paragraphs (3.1-4), we have Cotta's rebuttal (3.5-93). Here, however, due to lacunae in our text (3.64-65), a considerable part of Cotta's refutation, especially against the Stoic doctrine of providence (πρόνοια) has been lost.[9] Finally, 3.94-95 contains the conclusion of the dialogue, and to this I shall come back since it constitutes the main topic of this paper.

The dramatic date of the dialogue falls between 77 and 75 B.C.E.,[10] that is, at a time when Cicero is approximately twenty-nine to thirty-one years old. He has recently returned from Greece where he has pursued his philosophical studies.[11] He is therefore old enough and sufficiently trained to follow the discussion and to make up his mind as he does at 3.95.[12] Now the *De natura deorum* is quite unique among Cicero's philosophical works in that it is a dialogue neither in the manner of Heraclides Ponticus nor in that of Aristotle.[13] The former, according to Cicero, included as characters historical persons no longer living at the time of composition, whereas in the Aristotelian dialogue, again according to Cicero himself, the speeches of others are brought in in such a way as to give Aristotle

8. For what we know about C. Aurelius Cotta, C. Velleius, and Q. Lucilius Balbus, cf. Mayor, 1:xl–xlii; Pease 1:27–29.

9. The main lacuna is probably that in 3.64–65, but in all likelihood there are others as well. See Philippson, col. 1155.

10. Cf. Pease 1:24–26.

11. Cf. Reid, 3–5; Pease 1:25.

12. Pease, 1:24 states that at the time of the dramatic date of our work Cicero is a "youngish man, not expected to take an active part in the dialogue." But the reasons why Cicero does not take an active part in the conversation are, I believe, different, and more weighty than Pease assumes. See below.

13. On these two types of dialogue in Cicero cf. Pease 1:22–24.

the initiative.[14] A clear example of the Heraclidean type of dialogue is Cicero's *De republica*, of the Aristotelian, in the sense of having a contemporary setting,[15] the *De diuinatione*. The *De natura deorum* is different from both types of dialogue. Here Cicero is present in the discussion and old enough to take part in the conversation, as shown by the very fact that he expresses his opinion at the end. He is otherwise a real κωφὸν πρόσωπον and is practically ignored by the interlocutors.[16] Finally, it should be emphasized that throughout his philosophical dialogues Cicero pays attention to historical verisimilitude, in his characters and dramatic settings.[17]

In the *De natura deorum*, as in Cicero's other philosophical works, we find different kinds of inconsistencies. Some are due to misunderstanding and/or manipulation of his sources, others may be due to inconsistencies in the very sources he used.[18] Still other inconsistencies have to do with the ingrained habits of the writer who, forgetting that he is reporting a conversation, uses expressions such as "ut supra dixi" (2.65), "quam...supra diximus" (3.59), etc. None of these need be considered here. There is another type of inconsistency in the *De natura deorum* which has given rise to controversy. In 2.73 Balbus, referring to Velleius' statement in 1.18, says "uelut a te ipso hesterno die dictum est," while Cotta in 3.18 addresses the following words to Balbus: "omniaque quae a te nudius tertius dicta sunt, cum docere uelles deos esse." These references seem to imply that the whole discussion took up three days (one day for each book), and that there was an interval of one day between Balbus' speech in Book 2 and that of Cotta in Book 3. However, the work as extant seems to presuppose that the discussions occupying the three books took place on one and the same day. This follows not merely from there being one preface only for the whole work but more especially from the fact that the speech of every

14. Cf. *Ad Atticum* 12.19.4. But cf. also notes 15 and 50 below.
15. This limitation suffices for our purposes. Other questions concerning what Cicero means by the Aristotelian kind of dialogue are controversial and need not be discussed here.
16. For an exception cf. 2.104.
17. Cf. Ernst Becker, "Technik und Szenerie des ciceronischen Dialogs," (Dissertation, Münster, Osnabrück, 1938), esp. 11–30; and Pease 1:24 with n. 7; R.E. Jones, "Cicero's Accuracy of Characterization in his Dialogues," *American Journal of Philology* 60 (1939): 307–25. The latter publication, however, does not discuss the *De natura deorum*.
18. Cf., e.g., Mayor 3:ix ff.; L. Edelstein, *Studi italiani di filologia classica* 2 (1934): 131–83; Pease 1:26–27, 36 ff.

interlocutor follows immediately after the previous one without any intermission or new meeting whatever.[19] Some scholars have taken the two references to the fact that the discussions took place during three different days as relics of an earlier plan later abandoned,[20] but the inconsistency in time may also be due to hasty composition. However this may be, it is noteworthy that J.B. Mayor[21] found in the inconsistencies of time "conclusive proof that the book did not receive the finishing touches from the hand of its author." And from the wording of *De diuinatione* 2.3 where Cicero, after referring to the *Hortensius, Academica, De finibus,* and *Tusculanae disputationes*, says "Quibus rebus editis tres libri perfecti sunt de natura deorum," and in the *De fato* 2.1: "Quod autem in aliis libris feci, qui sunt de natura deorum, itemque in iis, quos de diuinatione edidi," Mayor infers that Cicero did not himself publish the *De natura deorum.* According to him, from the opposition of the words *editi* and *perfecti* in the former passage and of *feci* and *edidi* in the latter, one may infer that while the works referred to as *editi* had been published, the *De natura deorum* was what we should call ready for the press. Mayor concludes that if the *De natura deorum* was still unpublished at the time of Cicero's murder and if the manuscript was altered and emended by later editors or copyists this would go far to explain the roughnesses and inconsistencies in the dialogue. Mayor's thesis, however, accepted by Plasberg, Dougan, and Pohlenz,[22] among others, can hardly be right. For as Philippson points out,[23] how could Cicero refer to the *De natura deorum* three times in the *De diuinatione* (1.7-8; 2.3 and 148) and once in the *De fato* (1) if the work had not been published? Moreover, in the *De fato* (1) we read "itemque in iis, quos de diuinatione edidi." Since the *De diuinatione*, which is a continuation of the *De natura deorum*, was published, the *De natura deorum* must have been published too.

19. That the work as now extant has the discussions taking up one day only is shown by the following passages. 1.57: "Tum Cotta comiter ut solebat 'Atqui' inquit 'Uellei nisi tu aliquid dixisses, nihil sane ex me quidem audire potuisses....'" 2.1: "Quae cum Cotta dixisset, tum Uelleius...." 3.1: "Quae cum Balbus dixisset, tum adridens Cotta...." See also the device by which the dialogue is ended at 3.94: "Quae cum dixisset, Cotta finem. Lucilius autem 'Uehementius' inquit 'Cotta tu quidem inuectus es in eam Stoicorum rationem quae de prouidentia deorum ab illis sanctissume et prudentissume constituta est. Sed quoniam aduesperascit, dabis nobis diem aliquem ut contra ista dicamus....'"
20. Cf. Pease 1:16, n. 5 with the references given there.
21. Cf. Mayor 3:xxv–xxvi.
22. Cf. O. Plasberg, *M. Tulli Ciceronis De Natura Deorum* (Leipzig, 1917), iv; T.W. Dougan, *M. Tulli Ciceronis Tusculanarum Disputationum Libri Quinque* (Cambridge, 1905), 1:xvi ff.; M. Pohlenz, *Antikes Führertum* (Leipzig, 1934), 8.
23. Cf. Philippson, cols. 1151–52.

And we must infer that Cicero uses *editi, perfecti,* and *feci* for the sake of variety. In this connection one must call attention to the implications of the word *edere* in Latin, a translation of 'ἔκδοσις in Greek.[24] This is important in view of the conditions of ancient book making in general and for Cicero in particular, about whom there is abundant information in his correspondence with Atticus. "Publication" meant surely that the author himself had copies of his work made for distribution or that he permitted others to make a copy or copies. There were no royalties and no author's rights. In Cicero's case we know that he usually sent a copy to Atticus and to the dedicatee of a given work. In our case it is inconceivable that Cicero would not have sent a copy of the *De natura deorum* to Brutus. Atticus frequently had the work copied in his own house, though he was not a bookseller nor an editor of texts. In some instances we have evidence that Atticus received a copy of the work which later Cicero wanted to correct or substantially modify.[25] Such being the evidence about the "publication" of Cicero's works we may infer that they reached few readers and that it is highly improbable that Cicero did not "publish" the *De natura deorum.*

SINCE BOTH IN THE *De natura deorum* and elsewhere Cicero professes to belong to the Academy, readers from antiquity to the present have felt that the concluding statement of our work needs clarification. That statement itself must be discussed first, since it admits of two interpretations. In 3.95 Cicero, speaking for the first and only time since the introductory proem, says: "Haec cum essent dicta, ita discessimus ut Uelleio Cottae disputatio uerior, mihi Balbi ad ueritatis similitudinem uideretur esse propensior." This has generally and, I believe, correctly been taken to mean, "When these things had been said, we separated, the upshot being that to Velleius Cotta's argument seemed truer, whereas to me that of Balbus appeared more inclined to probability." It is grammatically possible, however, to take *Uelleio* not as a dative but as an ablative and to translate: "Cotta's argument seemed truer than that of Velleius, but to me that of Balbus seemed even more inclined to probability." A few critics seem to have so interpreted the clause,[26] but two objections against it are decisive: in the

24. Cf. B.A. van Groningen, "EKDOSIS," *Mnemosyne*, ser. 4, 16 (1963): 1–17.
25. For the evidence and a discussion of it cf. R. Sommer, "T. Pomponius Atticus und die Verbreitung von Ciceros Werken," *Hermes* 61 (1926): 389–422, an important paper whose results have not always been assimilated.
26. This may have been D. Hume's interpretation if it is right to infer (cf. Th. Zielinski, *Cicero im Wandel der Jahrhunderte,* 2nd ed. (Leipzig-Berlin, 1908), 286–87, with the note on p. 438), from the conclusion of his *Dialogues Concerning Natural Religion* (a work greatly influenced by the *De natura deorum*) that he took Uelleio to be an ablative: "so I confess, that, upon a serious review of the whole, I

first place, the word *mihi* is left without antithesis and its very position in the sentence becomes awkward; secondly, the philosophical differences between *uerior* and *ad ueritatis similitudinem propensior* are spoiled. In the generally accepted interpretation, Cicero uses these words to contrast the dogmatic attitude of Velleius to his own Academic probabilism.[27] The only possible advantage of the second interpretation is that Cicero would be passing judgment on all three speakers. But for Cicero the important issues are always those between the Stoa (or Stoic-influenced philosophers like Antiochus) and the New, skeptical Academy. He feels little sympathy for Epicureanism and sometimes he is not quite fair in his presentations of it. In this instance he probably considered that Cotta's lengthy refutation obviated any need on his part to reject it explicitly.

---

cannot but think, that Philo's principles are more probable than Demea's; but that those of Cleanthes approach still nearer to the truth." There are others also who seem, judging from their comments on 3.95, to have interpreted the sentence in the same way, e.g., R. Cudworth, *The True Intellectual System of the Universe...*to which are added the notes and dissertations of Dr. J.L. Mosheim, translated by J. Harrison, 3 vols. (London, 1845), 2:123, who apropos of 3.95 says: "more propense and inclinable to the doctrine of Balbus, than either that of Velleius or Cotta..., yet he did much prefer it (the Stoic doctrine) before, not only the Epicureanism of Velleius, but also the scepticism of Cotta." (Cudworth's work was first published in 1678). Cf. also Reid, p. 18, n. 4: "Cf. also the statements at the end of N.D. that the Stoic arguments about the gods are nearer truth than the rest."

27. Cf. Pease (1913), p. 27, who, in the formulation of the second objection, however, fails to point out that the ascription of both *uerior* and *ad ueritatis similitudinem propensior* to Cicero would make Cicero himself disregard the difference between the dogmatic assertion that truth is accessible to us and Academic probabilism. Note for example how the Stoic Quintus Cicero in *De diuinatione* 1.9 refers to 3.95 also from the point of view of a dogmatist: "satis enim defensa religio est in secundo libro a Lucilio, cuius disputatio tibi ipsi, ut in extremo tertio scribis, ad ueritatem est uisa propensior." Pease himself (1913), p. 26, n. 7, in this case remarks on the motivation of change from *ad ueritatis similitudinem* to *ad ueritatem.* His two remaining objections to the second interpretation of 3.95 are (i) to the brachylogy Uelleio Cottae disputatio uerior in the sense of *Uellei disputatione Cottae disputatio uerior* and (ii) to the fact that "instead of the comparative *propensior*, if three *disputationes* are involved, the superlative would be more natural." But in the case of (i) the brachylogy can be paralleled in Cicero, as he himself recognizes, and as for (ii), it seems to me that in *this* case the Academic probabilist Cicero would not have used the superlative.

Pease has sketched and argued against the two most common explanations of the purpose of Cicero's statement in 3.95.[28] The second interpretation is that, despite his being an Academic, Cicero is stating his true position when he declares Balbus' argument to be more probable than Cotta's. Since I believe that there is a kernel of truth in this interpretation, and since there is disagreement as to the exact nature and motivation of Cicero's siding with Balbus, I shall discuss it below.

The first interpretation is that in 3.95 Cicero utters a deliberate falsehood as he really agrees with Cotta's view, either because he does not wish to undermine the Roman state religion or because he fears an accusation of atheism. Because both of Cotta's speeches contain strong arguments against anthropomorphism and polytheism, the *De natura deorum* was very much used by the Latin Christian apologists in their polemics against paganism,[29] and they explicitly ascribed Cotta's position to Cicero himself. This is conspicuous in Arnobius' *Aduersus nationes,* where he praises Cicero's superior piety.[30] But if Arnobius does not mention Cicero's statement in 3.95, his student Lactantius does implicitly allude to it when he states "sed nimirum Socratis carcerem times ideoque patrocinium ueritatis suspicere non audes"[31] in connection with Cicero's refutation of Roman religion. And even M. van den Bruwaene, the author of the latest commentary on our work, offers a similar view of 3.95:

> On a cru pouvoir tirer de cette finale une indication sur l'opinion personnelle

28. Cf. Pease (1913), esp. 27–33 and Pease 1:33–35.
29. For a concise account cf. Pease 1:53–57 where the earlier bibliography is cited. Of later publications I mention only two: I. Opelt, "Ciceros Schrift *De natura deorum* bei den lateinischen Kirchenvätern," *Antike und Abendland* 12 (1966): 141–55; and H. Le Bonniec, "L'Exploitation apologétique par Arnobe du «De Natura Deorum» de Cicéron," in *Présence de Cicéron. Hommage au R.P. M. Testard* (Paris, 1984), 89–101.
30. Cf. esp. Arnobius, *Aduersus nationes* 3.6. Of 3.7 Alan Cameron, "Paganism and Literature in Late Fourth-Century Rome," in *Christianisme et formes littéraires de l'Antiquité tardive en Occident* (Vandoeuvres-Genève, 1976), p. 25, says that "there is a nice but doubtless apocryphal story in Arnobius that the senate once voted to have the *De natura deorum* burnt as a subversive work." Arnobius, however, merely states that he hears of some who think the Senate should have the *De natura deorum* destroyed. Cf. also Le Bonniec, "L'Exploitation," p. 90, n. 2. Some pagans probably did feel that the *De natura deorum* subverted paganism, and this may be the reason why Macrobius has Praetextatus say in the *Saturnalia* (1.24.4): "cum ipse Tullius, qui non minus professus est philosophandi studium quam loquendi, quotiens aut de natura deorum aut de fato aut de diuinatione disputat, gloriam, quam oratione conflauit, incondita rerum relatione minuat."
31. Lactantius, *Diu. Inst.* 2.3.5. On other related remarks cf. 1.17.4; 2.8.53; *De ira* 11.9; and Pease 1:34, nn. 1–2; p. 55 with nn. 7, 8, and 10; p. 56, with n. 1.

> de Cicéron. L'expression employée ici par lui est, en fait, assez extérieure: elle reprend simplement le slogan académicien de l'adhésion pratique motivéc par le plus probable. Elle est préparée par le rappel de Balbus que Rome doit son salut aux pratiques de la religion. Elle rend simplement à Cicéron sont étiquette de patriote académicien.[32]

Saint Augustine for his part explicitly states that Cicero made Cotta his mouthpiece in the *De natura deorum* and sided with Balbus because he did not dare say in his own name that there is no God.[33]

It is decisive against this kind of interpretation to recall that in the *De diuinatione*, the very continuation of our treatise, Cicero himself undertakes the attack against divination although he approved of divination for reasons of political expediency. This shows that it was not fear of appearing skeptical in regard to public religion that prompted him to avoid the role of Academic critic in the *De natura deorum.* Moreover, as Pease argues, if Cicero had really been afraid he need not have published at all, and he would not have revealed the different attitude of a Roman statesman in a public meeting and in a private conversation as he has Cotta do in 1.61. Nor could he have thought that the mere device of giving Cotta the role of the Academic skeptic would suffice to relieve him of responsibility for such views.[34] In addition, the conditions of publication in Cicero's case (referred to above) assured his philosophical works in particular of a small and select group of readers only; not the popular audience from which the alleged reaction could be expected.

32. Cf. M. van den Bruwaene, *Cicéron. De Natura Deorum. Livre III* (Brussels, 1981), 162, n. 397. Also other modern scholars refuse to take literally Cicero's statement in 3.95, e.g., L.F. Heindorf, *M. Tulli Ciceronis De Natura Deorum Libri Tres* (Leipzig, 1815), xii and 7; and others mentioned by A.B. Krische, *Die theologischen Lehren der Griechischen Denker* (Göttingen, 1840), 9, n. 1; and by Pease (1913), 27, n. 13.

33. Cf. *De Ciuitate Dei* 5.9 with 4.30. On the extent of the influence of the *De natura deorum* on Augustine's thought, M. Testard, *Saint Augustin et Cicéron* (Paris, 1958), 1:211–13, contends that it has been very great; while H. Hagendahl, *Augustine and the Latin Classics, Studia Graeca et Latina Gothoburgensia* 20.2 (Göteborg, 1967), 2:517–22, believes it played a subordinate role. In this case an intermediate solution seems more in agreement with the evidence.

34. Cf. Pease (1913), 29–30.

DISSATISFACTION with the two traditional interpretations prompted Pease to advance a third one of his own.[35] According to him, the purpose of the *De natura deorum* is not polemical or protreptic, as these interpretations assume, but rather descriptive. He believes that assuming a descriptive aim better explains the statement in the final sentence. Thus: "Cicero desires to give the impression of impartiality, which would not be produced by two Academics voting alike at the end. He also wishes to show to the reader an example of Academic method rather than of a dogma which might have been (even though wrongly) inferred from the consensus of two Academics, and to suggest that an Academic might use his individual liberty to select and accept any practical working principle, no matter from what school."[36]

Now I readily agree that Cicero's aim in the *De natura deorum* (and in many other of his philosophical works as well) is descriptive or more descriptive than polemical. But this has very little to do with the reason or reasons that led Cicero to make the statement in 3.95. Supposing for a moment that the work had ended without his final remark, a remark such as we find in no other work of his,[37] would the dialogue have lost its descriptive purpose? I believe not, because that purpose, to express it with Cicero's own words in the *De fato* (1), is

> Quod autem in aliis libris feci, qui sunt de natura deorum, itemque in iis, quos de diuinatione edidi, ut in utramque partem perpetua explicaretur oratio, quo facilius id a quoque probaretur, quod cuique maxime probabile uideretur, id in hac disputatione de fato casus quidam ne facerem impediuit.

That the discussion between Velleius and Cotta, on the one hand, and that

35. Cf. Pease (1913), 33–37 and Pease, 1:7; 28 and n. 7; 35–36.
36. Pease, 1:36.
37. There is no parallel at all in Cicero's other works. Pease (1913), 26, n. 5 apparently approves of Hirzel, *Der Dialog* (Leipzig, 1895), 1:533, who cites *De finibus* 5.95 as another instance where Cicero advances a positive opinion at the end of a work and does not remain a negative Academic. But to begin with, in the *De finibus* Cicero has the role of Academic critic throughout. Moreover, his remarks in *De finibus* 5.95, after Piso's exposition of Antiochus' doctrine and his admission that the Academics regard that system as inconsistent is nothing but Cicero's polite refusal to admit the doctrine while at the same time leaving the discussion open: "'Atqui iste locus est, Piso, tibi etiam atque etiam confirmandus,' inquam; 'quem si tenueris, non modo meum Ciceronem sed etiam me ipsum abducas licebit'." In fact, Hirzel's whole discussion of *De natura deorum* 3.95 (*Dialog,* 532–34) is vitiated by his notion that Cicero adopted the descriptive aim in the *De natura deorum* and later works because he had failed to gain converts to Academic skepticism with the earlier ones. For such a theory there is no evidence at all, and we must note that Cicero's technique in his dialogues is of a piece with what he conceives to be the essence of the Academic method of discussion. See above, 10 ff., with notes 56–57 below.

between Balbus and Cotta on the other are not considered settled or finished is shown by the fact that both Velleius and Balbus leave open the possibility of confuting Cotta's arguments against their respective philosophies.[38] This is a devise Cicero uses in other works as well[39] to indicate that the controversy is still viewed as open. Pease himself cites this statement of Bentley:

> ...that in all the disputes he (sc. Cicero) introduces between the various sects, after the speeches are ended every man sticks where he was before; not one convert is made (as is common in modern dialogue), nor brought over in the smallest article. For he avoided that violation of *decorum*; he had observed, in common life, that all persevered in their sects, and maintained every *nostrum* without reserve.[40]

In this statement it is not so much Bentley's explanation (which is at least controversial) as what he tries to explain — namely, that in Cicero's disputations the characters are not moved to change their opinions — that is important for our purpose here. For this fact is clearly related to Cicero's descriptive and objective aims: to set forth the ideas of the main schools of philosophy as objectively as possible and thereby to help the reader decide for himself.

The importance he assigned to his descriptive aim does not prevent Cicero from expressing his own views. He does so in the *De diuinatione* (1.7), a work closely related to the *De natura deorum* and having, according to his own words, the same descriptive purpose:

> Etenim nobismet ipsis quaerentibus quid sit de diuinatione iudicandum, quod a Carneade multa acute et copiose contra Stoicos disputata sint, uerentibusque ne temere uel falsae rei uel non satis cognitae adsentiamur, faciendum uidetur ut diligenter etiam atque etiam argumenta cum argumentis comparemus, ut fecimus in iis tribus libris quo de natura deorum scripsimus.

38. For Balbus cf. 3.94: "sed quoniam aduesperascit, dabis nobis diem aliquem ut contra ista dicamus." For Velleius cf. 2.1, esp. "sed ad ista alias," with Mayor's note *ad loc.*

39. Cf., e.g., *Academica Priora* 147–48; *De finibus* 5.95 cited in note 37 above.

40. *The Works of Richard Bentley,* ed. A. Dyce (London, 1838), 3:421; Pease, 1:36, n. 2.

The same thing may be said of the *De finibus*, conspicuous for its descriptive purpose and where Cicero does express his own opinion.[41] None of the reasons alleged by Pease can justify the conclusion that the descriptive purpose of *De natura deorum* presented to Cicero a special problem that resulted in his making the statement in 3.95. Clearly Cicero does not consider that advancing his own opinion puts undue pressure on his readers. In the very preface to the *De natura deorum*, when he complains about the excessive curiosity of those who desire to know his opinion, he also stresses the point that it is not so much weight of authority as force of argument that should be demanded and then promises to make up his mind about the question of the gods objectively and, in this case, regardless of his being an Academic.[42] It has been stressed above, and Pease himself recognizes it, that the *De natura deorum* is unique among Cicero's philosophical works in that, although he is old enough to follow the discussion and to make up his mind, he remains throughout the work a "silent character" (κωφὸν πρόσωπον) until 3.95. One must conclude that he chose this unusual setting for two related reasons: he wished *(a)* to give his opinion but *(b)* at the same time not to be one of the main speakers. In view of Cicero's practice in other works, such as the *De diuinatione*, Pease's theory fails to explain *(a)* and does not at all reckon with *(b)*.

Although Pease grants that Cicero's sympathies were divided between Stoicism and Skepticism, he fails to explain the statement in 3.95, and so is led to question either the accuracy or the sincerity of Cicero's words. Thus he contends that

> it is the Stoic proof, the discussion of Balbus, which, as Cotta says, "makes doubtful by its argument that which is in itself by no means doubtful (3.10, cf. 3.64)." To suppose, then, that he really accepts the Stoic *disputatio* is, I think, wrong; it is the positive convictions which lie behind it to which, "believing where he cannot prove," his assent is inclined.[43]

41. Cf. note 37 above.
42. Cf. 1.10; 1.14; 1.17.
43. Pease (1913), 36. Yet in other places and also in the introduction to his edition Pease himself stresses the fact that what Cicero accepts in 3.95 is Balbus' *Disputatio*. Also A.J. Kleywegt, "Ciceros Arbeitsweise im zweiten und dritten Buch der Schrift *De Natura Deorum*," (Dissertation, Leiden, Groningen, 1961), 219–21, rejects this and other of Pease's arguments. However, he does not relate 3.95 to the uniqueness of the dramatic setting and he still supposes that Cicero's main purpose was to produce a tie. But then Cicero need not have written 3.95 at all.

But why then did Cicero not say so? Why instead does he state that it is Balbus' *disputatio* that seems to him to be *ad ueritatis similitudinem propensior*? I shall come back below to this question and to Pease's objections against those who have taken Cicero's final statement more literally than he does. But it is possible to conclude even here that the descriptive aim of the *De natura deorum* is not directly related to Cicero's statement in 3.95. There is no reason not to take this statement literally, for unless Cicero had so meant it to be taken he need not have made it at all. Had Cicero said what he said in 3.95 merely or principally to balance Cotta's argument or opinions, he would purposely be making a false statement and would thus be violating the very principles of the Academic philosophy as he himself describes them in the preface to our work.[44]

THE NEXT INTERPRETATION to be discussed has been advanced by P. Levine.[45] It is not just an interpretation of Cicero's motivation in 3.95, but rather part of the author's theory that the *De natura deorum* as now extant is the result of substantial revision of the work by Cicero himself. According to Levine, Mayor's hypothesis cannot be readily explained away since to do so overlooks the unique character of the inconsistencies or irregularities of our work. These he sees not only in the incongruities of time referred to above but also in the fact that the *mise en scène* is the same for the conversations in the three books. For they are held the same day and at a single point without a break in time or change of setting. Such a dramatic structure is without parallel in any of Cicero's extant dialogues: nowhere does a single day extend beyond two books. Even more puzzling, according to Levine, is Cicero's inconsistency about himself between the introduction and the concluding statement. Thus, in the proem he complains about those who want to know his own opinion on each subject and "he intimates that he will suppress his private view so that his readers may not be unduly influenced in the formation of their judgement by the *auctoritas* of the writer."[46] Yet in the last sentence of the work he openly declares that he himself found greater probability in Balbus' presentation of

44. Cf. 1.1; 1.2; 1.7; 1.10; 1.11–12; 1.14; 1.17.
45. P. Levine, "The Original Design and the Publication of the *De Natura Deorum,*" *Harvard Studies in Classical Philology* 62 (1957): 7–36 (hereafter Levine[1]) and also the author's "Cicero and the Literary Dialogue," *Classical Journal* 53 (1958): 146–51, esp. 149–51 (hereafter Levine[2]).
46. Levine[2], 149.

Stoic theology. To these peculiarities are added *(a)* Cicero's peculiar role in the dialogue, where he remains silent throughout; and *(b)* the testimony of a letter to Atticus of June 45 B.C.E. where Levine sees evidence that Cicero originally intended to play a major role in the *De natura deorum*,[47] although he now appears in it as a mute Academic. He then argues that Cicero, susceptible as he was to public sentiment in a matter with such political implications as the attitude towards dogmatic theology, the very basis of all belief in the gods, decided "to extricate himself from the awkward position of being the *advocatus diaboli*."[48] In short, Levine's theory is that "the anachronisms and irregularities in the dramatic structure of the *De natura deorum* and the complete reversal of intention between the proem and the conclusion indicate something more than just hasty composition. They are the visible signs of an imperfectly executed change in the draft of the writer's original plan."[49]

There is no space in this paper to discuss in detail all the aspects of this ingenious theory, and it is my purpose to concentrate on its implications with regard to the statement in 3.95. I agree with Levine when he stresses the uniqueness of the dramatic setting and the unusual character of Cicero's role as a silent personage who at the end expresses his own opinion. But these peculiarities require in my view a different explanation from the hypothesis of a change of plan, which, *pace* Levine, is not supported by any external evidence. There is simply no indication either in the *De natura deorum* or anywhere else that Cicero ever intended to play a major role in this work.[50] In addition, the alleged

47. Cf. note 50 below.
48. Levine[2], 150.
49. Levine[2], 150.
50. In support of his theory that Cicero revised the *De natura deorum*, Levine adduces the following evidence: (i) *Ad Atticum* 13.19 of June 29, 45 B.C.E. From it he infers that Cicero preferred to have a prominent role in his later dialogues and that he departs from his practice in the *De natura deorum* and permits Cotta, the very person he had excluded from the *Academica*, to usurp the *principatus* (cf. Levine[1], 16–17). I agree that Cicero's role in the *De natura deorum* is peculiar and requires an explanation, but Levine's hypothesis is not warranted by the evidence. For the revised edition of the *Academica* Cicero reluctantly accepted Atticus' suggestion to include Varro but rejected Cotta because in that case he himself would have been a κωφὸν προσωπόν, something he is unwilling to do since the other character, Varro, is a contemporary. Moreover, in the *Academica* Cicero's role could not be different from that of Cotta (both being Academics) because the topic is concerned with theory of knowledge. The same is not the case in the *De natura deorum*. (ii) *Ad Atticum* 13.8 of June 9, 45 B.C.E., where Cicero asks Atticus to send him Brutus' abridgment of L. Caelius' *Annals* and Panaetius' Περὶ προνοίας. Levine maintains that of these two works almost certainly one and quite possibly both were used as sources for the *De natura deorum* (Levine[1], 17–18). But even

inconsistency between the final sentence and the proem is based on an unwarranted reading of the latter, as I hope to show.

Levine maintains that the final sentence

> requires further examination, for in one or two respects the informal poll of opinions that it contains is so curiously inconsistent with the rest of the dialogue that it appears to have been added as an afterthought to serve some immediate purpose rather than an integral part of the original plan.[51]

The first inconsistency detected concerns the Epicurean Velleius' siding with Cotta. For according to Levine, however much the Epicureans differed from the Stoics in their theology, they agreed with them that some form of divine existence is demonstrated by universal *consensus*. Hence Velleius' siding with Cotta would imply a somewhat unorthodox acceptance of the Academic denial of the validity of the argument from *consensus gentium*. If so, the sentiment attributed to Velleius at the end is not wholly consistent with the philosophic attitude which he meant to represent in the first book.

There are, however, two strong objections to this view. To begin with, as Levine himself admits, already in 3.65[52] there is an indication of the side Velleius is going to favor, and there is another such instance at the beginning of the third book (3.2):

---

if this is so, it does not follow that three weeks later, when he writes *Ad Atticum* 13.19 Cicero is already writing the *De natura deorum*. Moreover, both works may have been used — or Cicero may have intended to use them — for the *Tusculanae disputationes*. In any case, Caelius' work could have been used as a source for references to Roman history in any work whatsoever. As for Panaetius, he is mentioned only once in the *De natura deorum* (2.118), and his opinion is given in indirect discourse. Fatal to Levine's hypothesis is that when in 13.19.4 Cicero writes "quae autem his temporibus scripsi Ἀριστοτέλειον morem habent, in quo ita sermo inducitur ceterorum ut penes ipsum sit principatus," there is no reason to think that he in any way has the *De natura deorum* in mind. What is more, since in that passage he mentions the *De finibus* and the *Academica* but not the *Tusculanae disputationes,* and since the last mentioned work was written before the *De natura deorum* (cf. *De diuinatione* 2.3), the likelihood is that the sentence *quae...principatus,* if in it Cicero had in mind a work not mentioned by title, was meant to refer to the *Tusculanae disputationes*, in which Cicero has the *principatus* (cf. n. 71 below). Given the context of *quae...principatus* and the perfect tense *scripsi*, there is no reason to infer that Cicero was planning to have the *principatus* in the *De natura deorum*. Against Levine's theory cf. also K. Bringmann, *Untersuchungen zum späten Cicero, Hypomnemata* 29 (Göttingen, 1971), 266–68, who, however, does not comment on the two letters to Atticus discussed above in connection with Levine's thesis. He also fails to relate 3.95 to the peculiar dramatic setting of the work.

51. Levine[1], 18.

52. In 3.65, Velleius says to Cotta, who has asked permission for a detailed explanation of two topics, the world ruled by divine providence and providential care of human affairs, "Mihi uero...ualde uidetur; nam et maiora exspecto et is quae dicta sunt uehementer adsentior."

> Hic Uelleius "Nescis" inquit "quanta cum expectatione Cotta sim te auditurus. Iucundus enim Balbo nostro tuus contra Epicurum fuit; praebebo igitur ego me tibi uicissim attentum contra Stoicos auditorem. Spero enim te ut soles bene paratum uenire."

These two passages show that Velleius' role at the end of the dialogue is not merely an afterthought and/or oversight on Cicero's part. Moreover, the notorious rivalry between Epicureans and Stoics and the former's violent attacks upon the dogmas of Stoicism, prominent in Book 1,[53] are enough to justify, from Cicero's point of view, Velleius' siding with Cotta's attempt at refutation. Conversely, Balbus' approval of Cotta's criticism of Epicurean theology (on which Levine does not comment)[54] does not imply that he is slighting the cause of Stoicism. By these means and characterizations Cicero is conveying to us the idea that the rivalry between the two dogmatic schools caused each to welcome the negative dialectic that the skeptical Academy brought to bear against their respective dogmatic opponents.

Secondly, from a philosophical point of view neither Velleius nor Balbus is inconsistent in accepting Cotta's criticism of the rival school and at the same time subscribing to the argument from *consensus*. What is relevant is *not* the mere *consensus* about the existence of the gods but the fact that the Epicureans and the Stoics derive (at least in part) their peculiar and mutually conflicting conceptions of the nature of the divine from the argument from *consensus*. That this is the fundamental point for Cicero is shown by the very title of the work: *De natura deorum*.[55] Moreover, the epistemological foundations of the argument from *consensus* are quite different for Epicureans and Stoics. In short, then, in Velleius' siding with Cotta's argument against the Stoa there is no inconsistency with his role in Book 1.

The second inconsistency Levine sees in the concluding statement is as follows: Cicero involves himself in incongruity when, exercising his prerogative as an Academic, he says that he himself sees greater probability in the Stoic argument. For this disclosure is then contrary to his original intention as stated in the proem. Nowhere in the proem, however, does Cicero promise or imply that he will withhold his opinion. In 1.10 he

53. Cf. esp. 1.18–24 and 36–41.
54. Cf. 2.2: "Tum Balbus: 'Eundem equidem mallem audire Cottam, dum qua eloquentia falso deos sustulit eadem ueros inducat....Nam contra Epicurum satis superque dictum est'."
55. Cf. 1.2, esp. "quod uero maxime rem causamque continet, utrum nihil agant, nihil moliantur, omni curatione et administratione rerum uacent, an contra ab iis et a principio omnia facta et constituta sint et ad infinitum tempus regantur atque moueantur, in primis magna dissensio est."

complains only about the excessive curiosity of those who desire to know what his private opinion is; and, as said above, he stresses the point that it is not so much weight of authority as force of argument that should be demanded.[56] This is Cicero's usual way of emphasizing what he takes to be the characteristic Academic method and its purpose.[57] Moreover, in the proem itself of the *De natura deorum* Cicero promises to make up his mind on the question of the gods impartially and regardless of the fact that he belongs to the Academy.[58] Finally, it is noteworthy that the passage which Levine sees as an implicit promise by Cicero to withhold his opinion is part of a section of the proem that Pease with good reasons thinks "was written by Cicero without particular reference to the present work and later taken by him from that *volumen prooemiorum* from which, as he wrote to Atticus, he was in the habit of drawing when composing some treatise."[59] If this is so, it further reinforces my interpretation that 1.10 is part of a *general* defense of Academic method.

What is really necessary is to determine Cicero's attitude towards Stoic theology, something Levine does not attempt. In fact he states about Cicero's siding with Balbus: "Whether or not the statement reflects his own true sentiments on the subject is controversial, and the question need not be treated here."[60] Yet if the statement at the end of the dialogue truly reflects Cicero's opinion there is no need to suppose, as Levine and many other critics do, that Cicero ever planned to undertake the role of Cotta. In the next section, after some preliminary remarks, I shall present the evidence that Cicero accepted some important theological tenets of Stoicism, although from the point of view of a different epistemological foundation.

AMONG THOSE who have taken Cicero's statement in 3.95 to reflect his real views there has been disagreement as to its implications. Some scholars believe that he is an eclectic Academic with Stoic sympathies,[61] while

56. Cf. also notes 37, 42 and 44 above and 57 below with the corresponding remarks in the text.
57. Cf. also *Tusculanae disputationes* 5.11; 5.82–83; *De diuinatione* 2.150.
58. Cf. 1.17.
59. Pease 1:30. Cf. also 29–30 for the reasons in favor of considering 1.5–12 as an independent section containing a general defense of the Academic school and later used in the *De natura deorum*. For Cicero's *uolumen prooemiorum* cf. *Ad Atticum* 16.6.4; and Philippson, cols. 1127–28.
60. Levine[1], 21.
61. Cf., e.g., Cudworth, *System*, 121–23; F. Creuzer, *M. Tullii Ciceronis Libri Tres De Natura Deorum* (Leipzig, 1818), 693–94; Hirzel, *Dialog*, 532–34; A. Goedeckmeyer, *Die Geschichte des Griechischen Skeptizismus* (Leipzig, 1905), 150; Kleywegt, *Arbeitsweise*; etc.

others ascribe his siding with Balbus to his acceptance of the views of a particular philosopher, such as Posidonius or Antiochus.[62] Such interpretations, however, *(a)* fail to account for the evidence and *(b)* also meet strong objections. To deal with *(b)*, one must say, in the first place, that, whatever Cicero's immediate source (or sources) for Stoic theology, Balbus' exposition does not represent the peculiar views of any individual Stoic philosopher, nor those of the Middle Stoa or Antiochus. For the most part it draws on arguments that go back to the Old Stoa and which by and large can be considered the common property of the school. At times, Balbus tries even to represent, in addition to the common view, the dissenting conception of an individual philosopher, e.g. Panaetius.[63] Secondly, the term "eclectic" is certainly not an appropriate description of Cicero's philosophical attitude in general nor of that in our work in particular. He does not intend to build a philosophical system; he is merely an Academic probabilist who feels he can accept any doctrine of any school if it seems to him to be the most probable one, and who even reserves for himself the right to change his mind.[64] As to *(a)*, the interpretations described above do not account for the uniqueness of the dramatic setting of the *De natura deorum*, where Cicero is silent throughout but expresses his opinion at the end, and where because of his age and philosophical training he could have taken an active part in the discussion.

On the other hand, the objections brought by Pease against those who take literally Cicero's statement in 3.95 can be answered. His first objection is that "it is not the principles of the Stoics but the argument (*disputatio*) of Balbus which Cicero is said to consider the more probable."[65] But surely if Cicero believes, for example, that the Stoic argument from *consensus* is more probable than the Academic objections to it, then he surely accepts what the argument tries to prove, namely that there is some sort of supreme being that rules the universe. Of course for Cicero this would only be the "more probable view or opinion," not certainty and not knowledge. Pease's remaining objections amount to one main point, that: "the whole arrangement of the dialogue, with the advantage of the last word given to the Academic rather than to the Stoic, is an indication of the author's sympathies which cannot be entirely ignored."[66] Without repeating what was said above against Pease's interpretation, one must emphasize that his contention runs counter to Cicero's own estimate of what he

62. Cf., e.g., Hunt, *Humanism*, 131, with 134 and 137 (Posidonius).
63. Cf. 2.118.
64. Against Cicero as an eclectic cf. Reid, 13–15.
65. Pease 1:35.
66. Pease (1913), 32; cf. also Pease, 1:35.

accomplished in the *De natura deorum*. For in *De diuinatione* 1.7 and in *De fato* 1 he describes his method in our work as one in which he has carefully compared argument with argument so that it would be easier for anyone to approve what would seem to him to be most probable. Hence, at least Cicero did not consider that the Academic speaker's having the final word is an indication of his own sympathies or that the reader should be unduly influenced by that fact.[67] After all, Cotta's speeches are merely rebuttals of the two dogmatic schools and must therefore come after each of the dogmatic speeches. If Cicero had given Balbus the opportunity to answer Cotta, then Cotta also might say that he *could* answer Balbus and so on. What Cicero did was simply to indicate that the Stoic side considers it can answer the objections of Cotta, and so, that the issue is left open and not settled. And he probably considered that he had already fulfilled his purpose of describing for Latin readers the main tenets and arguments of the three schools — Epicurean, Stoic, and Academic — on the subject of the nature of the gods.[68]

The extant evidence does not of course suffice to determine in each instance what exactly were the points of Cicero's agreements and disagreements with the Stoic theology. But some can be established beyond reasonable doubt. However, it must be emphasized once more that in epistemology Cicero considers himself an Academic; he belongs to the Academy, at least as he takes the Academy to be: human beings cannot attain knowledge (the Stoics and Epicureans believe they can) because nothing in our presentations enables us to distinguish the true from the false. But human beings give consent to what seems to them the more probable view in each case, and that is enough to make social, political, and moral life worth living.[69] Thus he declares Balbus' *disputatio* to be not true, but *ad ueritatis similitudinem propensior*. This, then, is a fundamental difference between Cicero and the Stoics.

What Stoic arguments did Cicero in all likelihood find more probable than the corresponding Academic objections? I shall limit myself here to two main points. He admitted some kind of argument from *consensus* similar to the Stoic and not to the Epicurean one.[70] In the proem to the *De natura deorum* (1.2) he says: "Uelut in hac quaestione plerique (quod maxime ueri simile est et quo omnes [sese] duce natura uenimus) deos esse dixerunt."

67. Cf. references in note 56 above.
68. For this purpose cf. the whole proem to the *De natura deorum*, esp. 1.1–5 and 1.12–17.
69. Cf. 1.1; 1.11–12.
70. On the argument from *consensus* cf. R. Schian, *Untersuchungen über das 'argumentum e consensu omnium,'* Spudasmata 28 (Hildesheim-New York, 1973).

This opinion is independent of the dramatic purpose of our work, as is shown by parallel passages from other philosophical dialogues where there is no question that Cicero is speaking in his own name. Thus in *Tusculanae disputationes* 1.30 he states[71]:

> Ut porro firmissimum hoc adferri uidetur cur deos esse credamus, quod nulla gens tam fera, nemo omnium tam est inmanis, cuius mentem non imbuerit deorum opinio (multi de dis praua sentiunt; id enim uitioso more effici solet; omnes tamen esse uim et naturam diuinam arbitrantur, nec uero id conlocutio hominum aut consessus effecit, non institutis opinio est confirmata, non legibus; omni autem in re consensio omnium gentium lex naturae putanda est.

And again in the same work (1.36): "Sed ut deos esse natura opinamur quales sint ratione cognoscimus." The same opinion is forcefully stated in the *De legibus* (1.24-25):

> itaque ex tot generibus nullum est animal praeter hominem quod habeat notitiam aliquam dei, ipsisque in hominibus nulla gens est neque tam mansueta neque tam fera, quae non etiamsi ignoret qualem habere deum deceat, tamen habendum sciat. Ex quo efficitur illud ut is agnoscat deum, qui unde ortus sit quasi recordetur [agnoscat].

(The *De legibus* in fact contains several other Stoic doctrines, some of which are set forth in the second book of the *De natura deorum*.)[72]

It is also noteworthy that in the *De diuinatione*, just a few lines after having referred to the *De natura deorum*, Cicero approves of the argument from design (2.148): "et esse praestantem aliquam aeternamque naturam et eam suspiciendam admirandamque hominum generi pulchritudo mundi ordoque rerum caelestium cogit confiteri." And immediately afterwards links (true) religion with the knowledge of nature (2.149): "quam ob rem ut religio propaganda etiam est, quae est iuncta cum cognitione naturae," etc. This last is also a characteristically Stoic trait. The argument from design is also used by Cicero in *Tusculanae disputationes* 1.68-70, a passage too long to be quoted here.

71. Despite Pease 1:24, n. 3, who states that "The *Tusculans* are essentially timeless, with anonymous speakers...," it is a fact that the leading speaker is Cicero himself. Otherwise, how could the leading speaker, for example, cite Cicero's *Consolatio* as his own work (cf. *Tusculanae disputationes* 1.65–66)?

72. Cf. *De legibus* 1.18 ff. For example: Law is the highest reason implanted in nature (1.18–20); nature is governed by reason (1.21); man has reason implanted in him by the supreme god (1.22); reason is the common possession of man and god (1.23); the universe is a commonwealth of which god and man are members (1.23); etc.

Cicero's agreement with some of the doctrines of the Stoics would not be sufficient reason, however, for supposing that he could have undertaken Balbus' part. For, even apart from the fundamental epistemological differences between him and the Stoa, Cicero did not accept many fundamental Stoic doctrines. To give but one example, he could not have accepted the whole of Balbus' argument concerning the self-motion of the world-soul identified by the Stoics with the divine fire.[73] For the first of the *Tusculanae disputationes* shows that Cicero accepted the Platonic view of the soul as being perpetual self-motion but rejected the notion that it is of a fiery nature.[74]

What has been said is sufficient to show that it was not appropriate for Cicero to play a leading role in the *De natura deorum* whether as an Academic or as a Stoic. And so one must infer that the unique character of the dramatic setting of the *De natura deorum*, with Cicero present during the whole conversation and stating at the end that Balbus' argument seemed to him to be more probable than that of Cotta, was purposely chosen by him as fairer

73. Cf. 2.31–32.

74. Cf. *Tusculanae disputationes* 1.23 (on the different theories about the nature of the soul): "Harum sententiarum quae uera sit, deus aliqui uiderit; quae ueri simillima, magna quaestio est;" the argument of the *Phaedrus* is accepted in *Tusculanae disputationes* 1.53–54. In *Tusculanae disputationes* 1.60 Cicero says: "animae sit ignisne nescio, nec me pudet ut istos fateri nescire quod nesciam;" he adds that if Aristotle was right in introducing a fifth element, this must be the nature of gods and soul (*Tusculanae disputationes* 1.65); then, quoting from his own *Consolatio*, he rejects the view that the soul may be constituted by any of the four elements and asserts that there must be something peculiar belonging to the soul and different from the elements (*Tusculanae disputationes* 1.66): "Nihil enim est in animis mixtum atque concretum aut quod ex terra natum atque fictum esse uideatur, nihil ne aut humidum quidem aut flabile aut igneum." Cf. also *Tusculanae disputationes* 1.70: "Quae est ei [i.e., the soul's] natura? Propria, puto, et sua."

to his own position and to historical verisimilitude, always an important aim in the composition of his dialogues.[75]

A final point concerning the preceding discussion and Cicero's religious views in general. It is important to distinguish at least three things: *(a)* Cicero's attitude toward the Roman state religion; *(b)* his philosophical view based on probability that there is a rational power that governs the universe; *(c)* his real or nonexistent belief in a personal god or gods. Concerning *(a)* one may say that Cicero accepted the Roman state religion for reasons of political expediency as based on tradition. As for *(b)*, his opinion was based on rational belief on the ground of probability. But in the case of *(c)* it is important to remind ourselves that Cicero was not a Christian, nor had he been brought up in the Judaeo-Christian tradition. The question of monotheism versus polytheism is for him as for the ancients generally not a problem at all.[76] Nor was he a theist. Hence neither in the case of *(a)* nor in that of *(b)* would he have admitted a personal relationship with the divine nature. But whereas in the case of *(a)* the relation is essentially formal, in the case of *(b)* it is based on rational belief which does not exclude moments of doubt. In the most important questions, rational belief will not satisfy most men; but for some like Cicero, Socrates, or Plato it will always be the highest human achievement. Not because human reason is perfect but because it is our most precious possession.

■

75. Cf. note 17 above. I do not wish to deny that the dramatic setting and date Cicero chose for the *De natura deorum* had other advantages. They enabled him to avoid reference to recent political events and to emphasize the essential difference between the topic of the nature of the gods and the belief in divination (the *De diuinatione* has different speakers and there Cicero argues against the Stoic doctrine of divination). But neither of these (or similar) advantages are likely to have caused him to chose a dramatic setting where he remains silent during the whole conversation but expresses his opinion at the end.

76. The ancients may admit a highest god among the gods, but this is quite different from monotheism; what is essential in the monotheistic religions — Judaism, Christianity, Islam — is the polemical denial of any other gods apart from the only true god. On this subject cf. R. Pettazzoni, *Essays on the History of Religions,* Supplements to *Numen* 1 (Leiden, 1954): 1–10.

II
Augustus and Hadrian: Classical and Classicizing Modes

RICHARD BRILLIANT

BOTH AUGUSTUS AND HADRIAN were active patrons of the fine arts and architecture, the former concerned to give visual shape to the impartial idea, the latter to express his personal taste as an exercise of power. Through patronage and the acquisition of works of art, their eclectic preferences for Greek models differed in the direction and purpose of choice. Under Augustus, High Classical, especially Athenian, works were esteemed as paradigms of a new Golden Age: thus, the Ara Pacis Augustae, et Forum Augustum, and the adoption of the Polykleitan athletic male for the representation of the emperor, for his relatives, Marcellus, Gaius, and Lucius, and for the Genius Augusti. These and other revivalist choices contributed to the making of that moralizing, neoclassical art of Augustus, structurally coherent and convincingly synthetic, which transcended its artistic dependence and provided a fitting imperial mode for later Roman generations, including the nostalgic grecophil, Hadrian. Despite the profoundly hellenized character of his literary and artistic culture, Hadrian's allusive, even poetic invocation of the Greek part seems both artificial and slightly uneasy. The monuments of ancient and contemporary art and architecture, assembled for the emperor's delight in his Tivoli Villa, seem to assert too strongly the continuity and consistency of Greco-Roman civilization within the Empire, as if to deny the perceived decline of that civilization by an action of will. And yet, as revealed in his own image and in those of the too-supple Antinuous, a classicizing gloss was no substitute for that confidence in the Classical ethos, now lost.[1] Perhaps for Hadrian,

1. An earlier version of this essay was given at the Symposium on "Roman Patronage of Literature and the Arts," Columbia University, April 4, 1981. For the issues here see J.M.C. Toynbee, *The Hadrianic School: A Chapter in the History of Greek Art* (Cambridge, 1934); G. Rodenwaldt, *Kunst um Augustus* (Berlin, 1942); H. Jucker, *Vom Verhältnis der Römer zur bildenden Kunst der Griechen* (Frankfurt a.M., 1950); G. Traversari, *Aspetti formali della scultura neoclassica a Roma dal I al III sec. d.C.* (Rome, 1968); J.J. Pollitt, "The Impact of Greek Art on

the Greek past had become a glimmering mirage, no longer to be grasped in its essence.

In the Museum of Fine Arts, Boston, there is a handsome marble head of Augustus from Ariccia that appears to exemplify those ethical values, so prized by the Classical tradition and by the Augustan age [Fig. 1]. The large, clear features, quiet, even somber expression, restrained modelling, and youthful countenance convey an image of the noble prince who ruled the Empire of Rome. These formal elements are consistent with the normative patterns of Augustan portraiture, created early in his reign to project exactly such a grand, noble presence, dependent not on the reality of his changing appearance over the years but on the classical persona, developed specifically for this program of inspirational representation. The classicizing character of the Augustus portrait has long been recognized, but what is most interesting is the general scholarly belief that this head is a later work, either of the Claudian period or, more likely, of the time of Hadrian,[2] in both instances a revival of a form already overtly referential to fifth-century Greek idealizing models.

If the head in Boston is Hadrianic, it would provide another instance of the return to Augustan imagery and the replication of Augustan portraits, frequently encountered in Hadrianic art, a feature, perhaps, of this emperor's relentless pursuit of a normative classicism in whatever guise it could be found. Confirmation of this tendentious association appears in the marble portrait head of Hadrian's beloved Antinous, the supple Greek from Syria, now in the Staatliche Museum, East Berlin [Fig. 2].[3] The resemblance between these two heads strikes the eye and can be explained only by their common reference to an established Greek type and by the mutuality of their eclecticism. It is also possible that Augustan portraits served as prestigious and Roman intermediaries between Antinous, Hadrian's "beautiful

---

Rome," *Transactions and Proceedings of the American Philological Association* 108 (1978): 155–74; F. Coarelli and G. Sauron, "La Tête Pentini. Contribution à l'approche méthodologique du néo-atticisme," *Mélanges de l'École Française de Rome, Antiquité* 90 (1978): 705–51; H. Flasher, ed., *Le classicisme à Rome aux Iers siècles et après J.-C., Entretiens sur l'antiquité classique* 25 (Fondation Hardt, Geneva, 1979).

2. Claudian date, K. Vierneisel and P. Zanker, *Die Bildnisse des Augustus* (Munich, 1979), 69; Hadrianic, M.B. Comstock and C.C. Vermeule, *Sculpture in Stone: The Greek, Roman and Etruscan Collections of the Museum of Fine Arts, Boston* (Boston, 1976), no. 329.

3. Ch. W. Clairmont, *Die Bildnisse des Antinous: Ein Beitrag zur Porträtplastik under Kaiser Hadrian* (Rome, 1966), no. 16.

young man," and the noble young men of Greek fifth-century art, objects themselves of a more ideal form of admiration.

Indeed, the physical nature of Antinous' beauty must be distinguished even from the most intentionally beautiful of Augustus portraits, stemming from the Prima Porta statue in the Vatican [Fig. 7]. Prominent among them is the magnificent head of Augustus in Munich, wreathed in oak leaves, the *corona civica,* which symbolizes Augustus' rescue of the Roman state from the chaos of revolution and his imposition of peace and order [Fig. 3].[4] The transcendental quality of this image, carved in luminous Parian marble probably after 20 B.C.E., manifests a spiritual beauty of such purity that, although it has no formal precedent in Greek High Classical Art, the head seems to embody those abstract moral values in a most clarified state of representation. Some of Hadrian's early portraits, such as the cuirassed bust previously in Rome [Fig. 4], exhibit something of this refreshing confidence. At one level this portrait offers the comforting presence of the Emperor, unworn by care, and on another, signalled by the beard, the intellectual power of the Greek philosopher type, an admirable combination of Greek and Roman qualities and virtues in a convincing image of Hadrian, early in his reign.[5] Even if the cuirassed statue or bust is a Greek invention, it has been thoroughly Romanized,[6] by the beard æ and Hadrian was the first Roman emperor to wear one æ the beard was a familiar sign of the Greek and philhellene and, thus, a deliberate marker of Hadrian's personal inclination.

Such symbolic references, part of the corpus of signs like the himation which are culturally fixed, seem to function as a gloss on, rather than as the core of, the artwork. It is exactly the profound engagement in things "Greek," in the appropriation of the substance and not the surface of its models, and in the application of that repertory to cognate situations which characterizes Augustan art in its most sublime formulations. The posthumous statue of Marcellus as Hermes in the Louvre [Fig. 5][7] incorporates the cool, dignified forms of Greek classical statuary as a fitting vehicle for the portrait of the

4. Vierneisel and Zanker, *Die Bildnisse des Augustus,* 60, 61, 93; cf. head of Hadrian with the *corona civica* in a private collection, C. Isler-Kerenyi, "Hadrianos Olympios," *Antike Plastik* 15 (1975): 111–17, fig. 1.

5. In the Rome art market 1962, present whereabouts unknown. See M. Wegner, *Hadrian: Das römische Herrscherbild* (Berlin, 1956), 8 ff., for early portraits.

6. See K. Stemmer, *Untersuchungen zur Typologie, Chronologie und Ikonographie der Panzerstatuen* (Berlin, 1978), especially 131–48.

7. J.-Ch. Balty, "Notes d'iconographie julio-claudienne IV. M. Claudius Marcellus," *Antike Kunst* 20 (1977): 102–18.

noble son-in-law and putative heir of Augustus, dead too soon. The sculptor who signed himself, "Kleomenes, son of Kleomenes, the Athenian," would have known how to work in this manner because of his own classicizing training as an artist[8]; but the effort was not undertaken on his own initiative but, rather, because it suited the Classical values of his august patron.[9] As such, the Marcellus statue typifies Roman creative eclecticism, a form of knowing and complimentary imitation, motivated by a belief that the conditions that generated the original models are thought to have recurred under Augustus, thus legitimizing their appropriation and revival. The dividing line between authenticity and contrivance in such a charged situation is surely very fine.

> An imitation is related to the ancient models in two different ways: the first relationship is the natural result of being for a long time in close contact with the model and living with it, the second resembles it but results from the application of rhetorical rules. About the first kind there is little one can say, about the second one can say only that all the models have a natural grace and charm of their own, while their contrived imitations, even if they are perfect as imitations can be, always have something labored and unnatural about them.[10]
>
> Dionysius of Halicarnassus, *On Imitation* 7

Fully appreciative of the reflective analogy expressed in works of art like the Louvre Marcellus, Augustan patronage searched for a new synthesis, some ostensibly harmonious imagery which would reveal the tranquilized political situation, the *pax Romana,* and found it in Classical Greece. Yet, Augustus himself was no hellenist, he did not especially love Greece or the Greeks despite his study of the Stoic philosopher, Athenodorus, and he was not a collector of Greek art or of Greek artists, unlike many of his contemporaries and social equals, such as Asinius Pollio [Pliny, *Historia Naturalis* 36.23-24].[11] Nevertheless, Greek sculpture of the fifth and fourth centuries B.C.E., especially

8. See A. Giuliano, *La cultura artisitica delle provincie della Grecia in età romana* (Rome, 1965), 53 ff.; P. Graindor, *Athènes sous Auguste* (Cairo, 1927), 198 ff., 231 ff.
9. See E. Gabba, "Political and Cultural Aspects of the classicistic Revival in the Augustan Age," *Classical Antiquity* 1 (1982): 43–65.
10. G.M.A. Grube, *The Greek and Roman Critics* (Toronto, 1965), 211 ff.
11. The view of G. Rodenwaldt, *Kunst um Augustus,* passim; Suetonius, *The Life of Augustus.*

works associated with Athens in any way, seemed to establish the normative standards of classicism through which culturally ideal forms of expression had been and could continue to be expressed.[12]

For the image of Augustus and his male relatives, artists resurrected the Polykleitan model both as a faithful replica and as a creative adaptation. The Naples Doryphorus [Fig. 6] comes from the palestra in Pompeii and appears to be an academic copy of the bronze original, probably made during the reign of Tiberius.[13] Although not the original, the statue is an an authentic copy of a famous work of Polykleitos, master of the Classical male nude, who embodied in the Doryphorus his aesthetic theories of ideal proportions and created, in effect, the acknowledged image of High Classical art. The popularity of the Polykleitan mode under Augustus, and to a lesser degree under Hadrian, must be seen both as an expression of Roman cultivated taste and as an exploitation of its image value for purposes of idealization.[14]

Modernized Roman copies of Greek ideal, or idealized, statuary, constituted a major corpus of artistic achievement beginning in the late Republic.[15] The Polykleitan model was frequently employed to represent the *Jugendromantik* of the Augustan age, and thus participated fully in Golden Age imagery given full expression by the regime. This model also served to define the very idea of the Emperor and to shape the Prima Porta statue of Augustus [Fig. 7], perhaps the most famous and most complex cuirass statue of a victorious commander, surviving from antiquity.[16] Although the Prima Porta Augustus may itself be a Tiberian copy of a bronze original, created after 27 B.C.E., the statue manifests the very essence of Augustan classicism, a millenarian ideology quite visible.

> For other peoples will, I do not doubt
> still cast their bronze to breathe with softer features,
> or draw out of the marble living lines,
> plead causes better, trace the ways of heaven

12. See especially P. Zanker, "Zur Funktion und Bedeutung griechischer Skulptur in der Römerzeit," in *Le classicisme à Rome,* 283 ff.; and his comment on an earlier paper, reported on p. 45.
13. P. Zanker, *Klassizistische Statuen: Studien zur Veränderung des Kunstgeschmacks in der römischen Kaiserzeit* (Mainz, 1974), 7 ff.
14. See G. Gullini, *Il «classicismo» Augusteo: Cultura di regime* (Rome, 1978).
15. See R. Wünsche, "Der Jüngling vom Magdalensberg. Studie zur römischen Idealplastik," in *Festschrift Luitpold Dussler* (Vienna, 1972), 45–80.
16. H. Kähler, *Die Augustusstatue von Primaporta* (Cologne, 1959); F. Johansen, "Le portrait d'Auguste de Prima Porta et sa datation," in *Studia Romana in honorem Petri Krarup* (Odense, 1976), 49–57; H. Jucker, "Dokumentation zur Augustusstatue von Primaporta," *Hefte des Archäologischen Seminars der Universität Bern* 3 (1977): 16–37.

with wands and tell the rising constellations;
but yours will be the rulership of nations,
remember, Roman, these will be your arts:
to teach the ways of peace to those you conquer,
to spare defeated peoples, tame the proud.[17]

Virgil, *Aeneid* 6.847-853

Despite Virgil's disclaimer, Greek artistic and ethical concepts had penetrated beneath the carapace of Roman martial virtue, thereby contributing substantially to the transcendental character of Augustus in his statue from Prima Porta. For the Augustan image-makers Classical Greek art was instrumental, a means of achieving an authoritative public presence through works of art which possessed the requisite conventions of nobility. In such monuments the Greek artistic tradition remained alive and effective because its value system was essentially intact, even when adapted by Imperial Romans.

For Hadrian, whose charming, wandering soul is given to jokes and play (Hadrian, *Poet.* 3),[18] the classical tradition is intrinsic to his personality. He is truly the heir to Tiberius, a classicist by training and inclination, looking to the past for values no longer found, indulging in nostalgia for a dying world, alienated from his declining culture, a Sophist in the arts who, at his Tivoli Villa, is the collector par excellence of the worthy remnants of Greco-Roman art for his private delectation.[19] Hadrian's personal approach to Greek art, his assumption of a classical manner in the exercise of patronage seem to be the last refuge of a sentimental idealist. Hadrian took pleasure in recollection [*Fuit memoriae ingentis*] and contemplative reverie, while seeking to restore his wandering soul through art like some Alexandrian intellectual, looking back to the Golden Age of Athens.[20]

For many years the statue of Hermes in the Vatican was called the "Belvedere Antinous" [Fig. 8] because scholars ands amateurs saw it as the full realization of Hadrian's taste in objects of art and of love.[21]

17. A. Mandelbaum, *The Aeneid of Virgil* (Berkeley, 1981), 166, 167.
18. *Scriptores Historiae Augustae,* "Hadrian" 25.9; note I.5, which notes that Hadrian was so devoted to Greek studies that some called him "Graeculus."
19. Views well expressed by M. Yourcenar, *Memoirs of Hadrian* (New York, 1954); E. Clark, *Rome and a Villa* (New York, 1962).
20. See Toynbee, *The Hadrianic School,* xiii ff.; Graindor, *Athènes sous Hadrian,* 37 ff., 197 ff., 213–84 (Hadrian "a new Pericles").
21. W. Helbig, *Führer durch die öffentlichen Sammlungen klassischer Altertümer in Rom,* 4th ed. (Tübingen, 1963), 1, no. 246.

Now recognized as a Hadrianic copy of a Greek bronze original of the fourth century B.C.E., the elegant, sinuous figure of the god, Hermes, represents a late classical image of male beauty, perhaps derived from Praxiteles, a work that seems simultaneously approachable, as a human body, and remote, as an ideal object of art. The aesthetic attitude toward such works had already been expressed by the Sophist, Dio Chrysostom, in his *XXI Discourse: On Beauty,* composed in the late first century A.D., when Hadrian was a young man:

> How majestic the youth is an handsome; and what is more, his appearance is ancient or classic (*archaion*) in type, such as I have not seen in your modern statues, but only in those set up at Olympia, the very old ones. The images of the subsequent periods even show a steady decline.... Masculine beauty has [declined].... And if people do by any chance take an interest in handsome men, it is in a wanton way and for no good purpose.[22]

Thus, the ground was laid for Antinous, whose idealizing, sentimental images — both *kouros* and *eratos* — emerged from an eclectic pastiche of Greek models, assimilated in a hellenistic manner. The prospective comparison between the Vatican Hermes [Fig. 8] and images of Antinous [Fig. 9] is especially apt. The grave countenance of Hermes suggests his role as a psychopomp, the transporter of the souls of the dead to the other world, while the sculpted portraits of Antinous appear to date from after his death and deification in Egypt in 130 A.D. and so express Hadrian's profound nostalgia for his lost love in the last eight years of his life.[23] Antinous' marble statue in Delphi [Fig. 9] may be the closest in sculptural quality and erotic characterization to the lost prototype, presumably created for Hadrian,[24] and subsequently disseminated through replication and adaptation, just like some Greek masterwork. Now Antinous is forever young, forever beautiful, a unique object of an emperor's overmastering love in the Greek fashion:

> Holy Socrates, why always with deference
> Do you treat this young man? Don't you know
> greater things?

22. J.W. Cohoon, Loeb Library ed. (Cambridge, MA, 1980), 2:271 ff.
23. See Clairmont, *Die Bildnisse des Antinous,* 13–21.
24. Ibid., 21 ff.

Why so lovingly, raptly,
As on gods, do you gaze on him?
Who the deepest has thought loves what is most alive,
Wide experience may well turn to what's best in youth,
And the wise in the end will
Often bow to the beautiful.

Hölderlin, *Socrates and Alcibiades* (1798/9)[25]

In 123 C.E. Hadrian visited Pergamon, an important stop on his continual tours of the Empire. To commemorate his visit a nude statue was placed in the sanctuary of Asklepius [Fig. 10],[26] a statue that combines a familiar Greek heroic type, signs of the Imperator temporarily disarmed, and the haunting, bearded face of the intellectual. Romans had gradually become accustomed to heroic nude statues of their emperors, although his appearance in the toga or in the cuirass [Figs. 4, 7] was preferred, and Hadrian did not hesitate to employ the rich, established repertory of Imperial iconography. However, only Nero and Domitian before him had displayed their fascination with Greek culture in their public behavior, but no Roman emperor had ever exhibited that personal preference in works of art for private and public use. The master of the Antinous portrait had responded to Hadrian's private need; it was the emperor who publicized that private, intimate experience through the multiplication of his beloved's images. Thus, Antinous, a Greek and beloved of a Greekling, became detached from Hadrian's memory through the retrospective invocation of Greek art.

Augustus had remade the ancient world. Greece was an integral part of that world, and he did not hesitate to incorporate references to Greek art in his greatest public monuments, such as the Ara Pacis and the Forum Augustum. In the latter, his artists replicated the caryatids from the Erectheoin on the Athenian Acropolis but converted them to Roman service and subordination.[27] In his turn, Hadrian copied the same caryatids more accurately and displayed them together with many other replicas of ancient art in the Canopus of his Villa at Tivoli,[28] as part of his personal museum of objects and memories.

25. F. Hölderlin, *Poems and Fragments,* trans. M. Hamburger, rev. ed. (Cambridge, 1980). 67.
26. Wegner, *Hadrian*, 105.
27. See P. Zanker, *Forum Augustum* (Tübingen, 1968), 12 ff., figs. 25, 26; R. Brilliant, *Art Bulletin* 53 (1971): 110–13.
28. See A. Aurigemma, *Villa Adriana* (Rome, 1961), 108 ff., figs. 95 ff., pls. VI, VII; J. Raeder, *Die statuarische Ausstattung der Villa Hadriana bei Tivoli* (Frankfurt a.M., 1983), 213 ff.

> His villa at Tibur was marvelously constructed, and he actually gave parts of it the names of provinces and places of the greatest renown, calling them, for instance, Lyceum, Academia, Prytaneum, Canopus, Poecile, and Tempe. And in order not to omit anything, he even made a Hades.[29]
>
> *Scriptores Historiae Augustae,* "Hadrian," 26.5

■

29. Translated by D. Magie, Loeb Library ed. (Cambridge, MA, 1953), 1:79.

Fig. 1. Augustus. Museum of Fine Arts, Boston. Museum photo.

Fig. 2. Antinous. Staatliche Museen, Berlin. Museum photo.

Fig. 3. Augustus with *corona civica*. Glyptothek, Munich. Koppermann photo.

Fig. 4. Hadrian. Formerly Roman art market, 1962.

Fig. 5. Marcellus Hermes. Louvre, Paris. Alinari 22668.

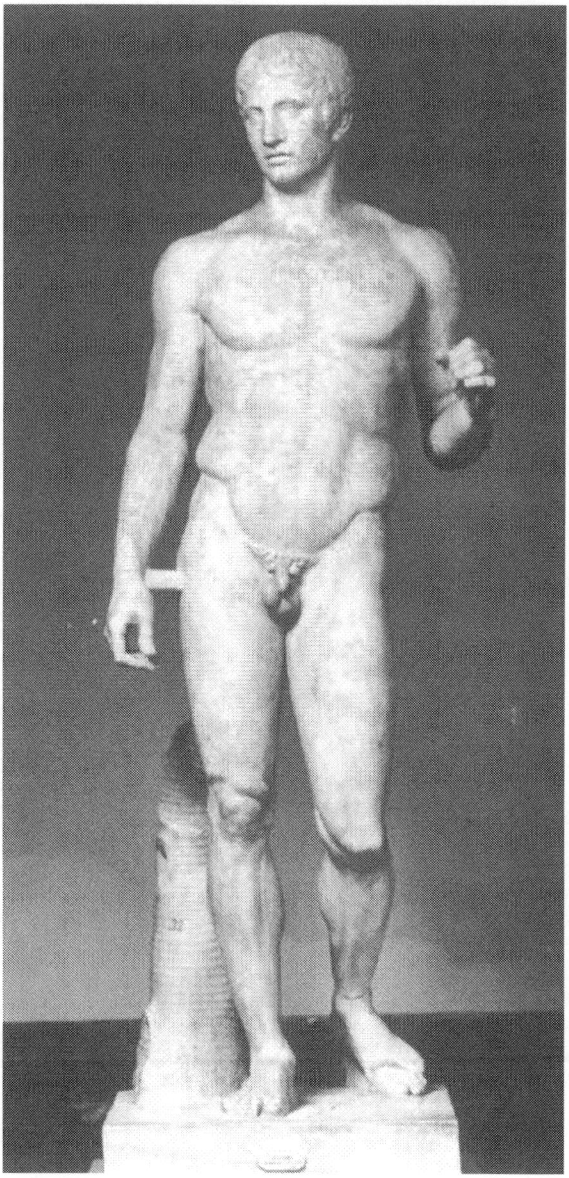

Fig. 6. Doryphorus. Roman copy of Polykleitan original. Museo Nazionale Archeologico, Naples. Hirmer photo.

Fig. 7. Prima Porta Augustus. Vatican Museum, Vatican City. Museum photo.

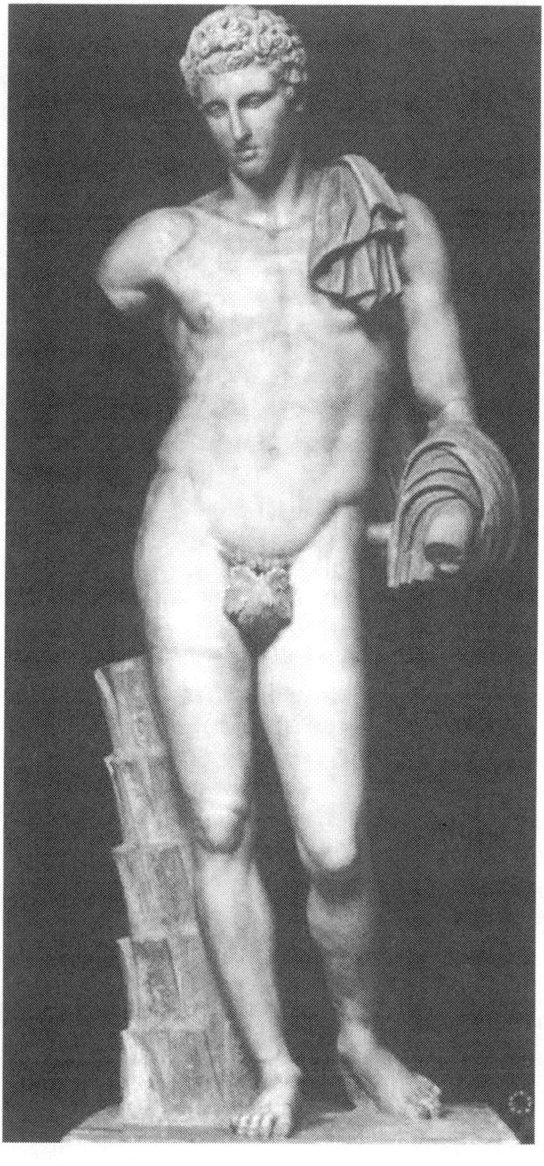

Fig. 8. Hermes. Formerly the "Belvedere Antinuos." Vatican Museum, Vatican City. Anderson 1303.

Fig. 9. Antinuos. Delphi. Museum photo.

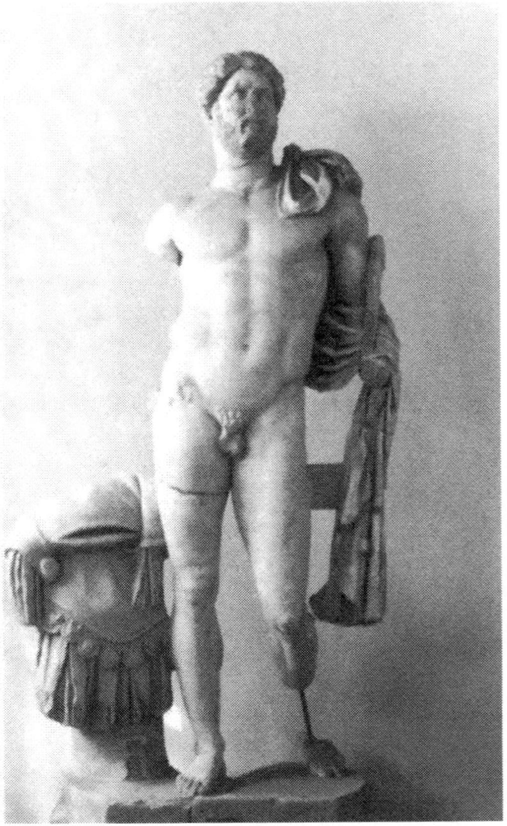

Fig. 10. Hadrian. Pergamon. Museum photo.

III
Church, State and Divorce in Late Roman Egypt

ROGER S. BAGNALL

A RECENTLY PUBLISHED PAPYRUS from Oxyrhynchos[1]; contains a petition from one Aurelia Attiaina; to Flavius Marcellus; tribune and officer in charge of the peace.[2] She recounts that

> A certain Paul, coming from the same city, behaving recklessly carried me off by force and compulsion and cohabited with me in marriage. [She had a female child by him and lived with him in her house. He behaved badly (the context is damaged) and cohabited with another woman. Then] after some time again he beguiled me through priests until I should again take him into our house, agreeing in writing that the marriage was abiding and that if he wished to indulge in the same vile behaviour he would forfeit two ounces of gold, and his father stood surety for him. I took him into our house, and he tried to behave in a way that was worse than his first misdeeds, scorning my orphan state, not only in that he ravaged my house but when soldiers were billeted in my house he robbed them and fled, and I endured insults and punishments to within an inch of my life.
>
> So taking care lest I again run such risks on account of him, I sent him through the *tabularius* a deed of divorce through the *tabularius* of the city, in accordance with imperial law. Once more behaving recklessly, and having his woman in his house, he brought with him a crowd of lawless men and carried me off and shut me up in his house for not a few days. When I became

1. *P.Oxy.* L 3581

2. This paper grows out of a portion of a lecture given at Bar-Ilan University (Israel) and at the Hebrew University of Jerusalem in May, 1986. I am grateful to Ranon Katzoff for the invitation to Bar-Ilan, which prompted its writing, and to the audiences at both institutions for their comments.

> pregnant, he abandoned me once more and cohabited with his same so-called wife and now tells me he will stir up malice against me. Wherefore I appeal to my lord's staunchness to order him to appear in court and have exacted from him the two ounces of gold in accordance with his written agreement together with such damages as I suffered on his account and that he should be punished for his outrages against me.

Scarcely an edifying tale. The document has, alas, lost its dating formula, and the editor dates it to the fourth or fifth century on the basis of the handwriting.[3] As the editor points out, such a date makes the text of interest for the question of the interaction of church and state in the restricting of divorce in late antiquity. That in the fourth to sixth centuries both civil and ecclesiastical authorities sought to discourage divorce has long been known; but the nature of these attempts, their relationship to one another, and their effects in practice have all, I think, been misunderstood.

First, the civil restrictions. The Roman law of marriage had for centuries allowed both divorce by mutual consent and unilateral declarations ending marriages, with hardly any restrictions,[4] until Constantine; introduced new regulations in 331 (*CTh* 3.16.1): women were forbidden to send notices of repudiation to their husbands on "slender" grounds such as that they were drunkards, or gamblers, or philanderers. Wives could use only homicide, sorcery, or tomb destroying as grounds; otherwise, they would be deprived of property and exiled to an island. A man could claim only adultery, sorcery, or procuring, on penalty otherwise of restoring the wife's dowry and not being able to remarry — and if he did remarry, his former wife could seize his house and the dowry of the new wife. None of the restrictions thus actually invalidated either the divorce or the new marriage.

Constantine's legislation lasted three decades. Julian; seems, to judge from a passage of 'Ambrosiarter', to have cancelled Constantine's penalties for forbidden repudiation[5]: "Ante Iuliani edictum, mulieres viros suos dimittere nequibant. Accepta autem potestate, coeperunt

3. The reference to priests, *presbyteroi*, guarantees in any case a date after the church in Egypt came above ground in the second decade of the fourth century, after the edict of toleration of 313.
4. F. Schulz, *Classical Roman Law* (Oxford, 1951), 134, notes that Augustus forbade remarriage by a freedwoman who divorced without his consent a husband who was also her patron.
5. See M. Kaser, *Das Römisches Privatrecht,* 2 ed. (Munich, 1971), 175 ff. The passage is Augustine, *Quaestiones de utroque testamento mixtim* 115 (PL 35: 2348–49).

facere quod prius facere non poterant; coeperunt enim quotidie viros suos licenter dimittere." One fragment of this law, dated 363, apparently survives in theTheodosian Code,[6] but we do not have its text for the part which concerns us here. Augustine;, to be sure, misleads in suggesting that Julian changed a situation existing from time immemorial, rather than merely restoring longstanding Roman custom, but his purpose in this argument did not make exact legal history necessary or, perhaps, desirable.

So far as we know, it was not until 421, long after Julian's death,; that restrictions were again imposed. In *CTh* 3.16.2, Honorius; imposed a new set of distinctions. Repudiation by the woman without cause led to the loss of marriage gifts and dowry and to her deportation, as well as to denial of the right of remarriage; repudiation for ordinary traits of bad character led to the same financial penalties and denial of remarriage, but not to deportation; and repudiation for serious crimes allowed her to remarry after five years had passed. Husbands' repudiations were divided in the same three classes, but with somewhat different penalties: groundlessly, the same penalties as for the woman who divorced on grounds of bad character (and the repudiated woman would be allowed to remarry after a year); on bad character, he returns the dowry, recovers his gifts, and can remarry after two years; on grounds of criminal behavior, he keeps all of the property and can remarry immediately.

Eighteen years later, Theodosius; (in *NovTheod* 12, 10 July 439) ordered that the dissolution of marriage required the sending of a *repudium*.[7] He went on, however, to say, "But in sending a notice of divorce and in investigating the blame for a divorce, it is harsh to exceed the regulations of the ancient laws. Therefore the constitutions shall be abrogated which commanded now the husband, now the wife to be punished by the most severe penalties when a marriage was dissolved, and by this our constitution we decree that the blame for divorce and the punishments for such blame shall be recalled to the

6. *CTh* 3.13.2 (363); this passage deals with dowry and marital gifts.

7. The reference to imperial law in *P.Oxy.* 3581 is clearly tied to the notion of *repudium*: "I sent him through the *tabularius* a *repudium* through the *tabularius* of the city in accordance with imperial law." This reference seems to me to refer most obviously to the requirement of a *repudium* stated in the legislation of 439 and repeated in 449. (Coles in his note to line 16 cites *P.Stras.* III 142 [a divorce document of 391, discussed below] as evidence for Julian's repeal of Constantine's legislation, thus allowing a fourth-century date for this papyrus. But, as we shall see, the Strasbourg papyrus reflects a divorce by mutual consent, not affected by either Constantine or Julian, and thus is irrelevant to the date of the Oxyrhynchos papyrus.)

ancient laws and the responses of the jurisprudents." This law of Theodosius; followed by little more than a year the appearance of his Code, which included the law of Honorius; (issued in the name of Theodosius and Constantius; as well as himself) quoted above, which it appears to contradict.

Ten years later, however, Theodosius; (*CJ* 5.17.8) returned to a more specific regulation of the causes for divorce which could be cited in a *repudium*. This time the sexes are treated more equally, for a woman has a wide range of causes allowed, including the husband's adultery, criminality, or violence toward her. Justified repudiation allowed the woman to recover her property and remarry after a year, a man to remarry immediately. Unjustified repudiation subjected the woman to a five-years' wait, the man to forfeiture of the dowry. Anastasius later (*CJ* 5.19.7, of 497) allowed the woman to remarry after a year if a groundless repudiation was by common consent.[8]

These meanderings of imperial legislation have received quite differing interpretations from modern scholars. It is neither possible nor useful to cite all of the extensive bibliography here; two examples of imprecision by great scholars will suffice. H.J. Wolff [9] describes Constantine's legislation as a "rigid restriction of justified divorce to a few cases of grave offense," and regards the legislation of 421 and 449 as a continuation of its development, while the novel of 439 is a "shortlived return to classical principles." A.H.M. Jones[10] claims that "since divorce under the old legal forms had been rendered so difficult, many couples dissolved their marriage by consent. This was forbidden by Theodosius II in 439, but he at the same time abolished all the penalties for divorce, and went back to the classical law." It is true, of course, that not requiring grounds for termination of the marriage is classical, but the requirement of a formal *repudium* to accomplish this termination is not. On the other hand, it is surely incorrect to think that this requirement made divorce by mutual consent impossible. Anastasius' constitution, mentioned above, shows that that is not what

8. A.H.M. Jones, *The Later Roman Empire, 284–602* (Oxford, 1964), 975, (= Jones, *LRE*) wrongly describes this law as ruling "that if a husband divorced his wife with her consent" she could marry after a year. Anastasius actually does not distinguish the sender of the repudiation, ordering only that it be sent by common consent ("Si constante matrimonio communi consensu *tam mariti quam mulieris* repudium sit missum, quo nulla causa continetur").
9. "Doctrinal Trends in Post-Classical Roman Marriage Law," *Zeitschrift der Savigny-Stiftung für Rechtsgeschichte, Rom. Abt.* 67 (1950): 262. This article, occupying 261–319, has very extensive bibliographical notes.
10. *LRE*, 974.

he thought the law meant; and Valentinian's abrogation of the Theodosian novel of 439 (*NovVal* 35.11, of 452) suggests that that novel's burden for him was its permission for unilateral and unjustified repudiation, not any restrictions put on consensual divorce.[11] Indeed, Justinian's legislation in the next century prohibiting divorce by mutual consent (*Novel* 117.10) says explicitly that it was not previously prohibited.

To summarize the course of legislation: apparently Constantine's limits on the possible grounds for unilateral divorce were repealed by Julian some thirty years later, and for six decades there was no imperial legislation known to us on the subject. In 421 similar restrictions were reintroduced; in 439 they were removed in the East in favor of a requirement for written repudiation, a law accepted in the West in 448[12]; but in 449 Theodosius changed his mind and again reintroduced restrictions (a bit more systematic and equitable this time), a course essentially followed by Valentinian in 452 with his return to the law of 421. Apart from Anastasius's clarification late in the century, the next major change did not come until Justinian tried to prohibit divorce by mutual consent in 542, a ban repealed by Justin in 566 (*Novel* 140).

Many modern works, apart from failing to observe consistently the distinction between unilateral and consensual divorce and that between prohibiting and penalizing a divorce, go on to describe this legislation as the result of Christian influence. "Christian legislation allowed divorce only for certain reasons and instituted penalties in case of contravention," said Schulz.[13] "The Christian empire, however, brought restrictions on divorce itself," said Thomas.[14] Such statements will not do; the imperial legislation of the fourth and fifth centuries is not "Christian" legislation, and the empire was not a "Christian" empire, no matter how often it is called one. But underlying these remarks is a serious debate about the extent of Christian influence on the imperial legislation on divorce. Positions have ranged from the view expressed by Schulz and Thomas to the opposite extreme, a denial

11. Valentinian's novel orders that "those regulations which were decreed by our sainted father Constantius shall be preserved inviolate." Pharr's footnote: "Not extant, but cf. *CTh* 3, 16, 1." An error for citing 3.16.2? The latter law includes Constantius among the Augusti, in 421, and is surely what is meant.

12. As part of the general affirmation of Theodosius' laws after the Code, see *NovVal* 26.

13. Above (n. 3), 134.

14. *Textbook of Roman Law* (Amsterdam, 1976), 426.

of all Christian influence.[15] Jones rather enigmatically remarked that "the Christian emperors tightened up the laws of divorce, but not in an entirely Christian sense."[16]

There is no doubt that the emperors of the fourth and fifth centuries did enact some laws which show a distinctively Christian viewpoint. A random example is the edict of Valentinian, Theodosius, and Arcadius (*CTh* 9.7.5) in 388 forbidding any marriage of a Jew with a Christian; such a marriage is to be considered adultery, and anybody is allowed to accuse the culprits. Many others could be cited. Nor was the church shy about asking for imperial legislation when it thought it needed; the title on heretics in the Theodosian Code (*CTh* 16.5) is sufficient evidence.[17] And there is even one piece of well-known evidence for a request by a church council (that of Carthage in 407) for imperial legislation to prohibit remarriage while a former spouse is alive[18]; it seems to have been ignored. On the other hand, the church showed considerable reluctance to try to impose generally on this world what it proclaimed as the ethical norms of the City of God.[19]

But these are not sufficient considerations to establish the intent of these laws. For that, two important points need examination. First, do the laws correspond to the position of the church both on divorce and on the relationship of church and state with respect to it? And second, what do the legislators have to say about their reasons?

The church took a strong position against divorce right from the start. Paul (1 Cor 7:8-11) gave as Christ's prescription that those who were married should not be separated and that a woman should not leave her husband; if she did, she should reconcile with him or stay unmarried; and that a husband should not dismiss his wife. This position rests on the same tradition reported in the Synoptic Gospels

15. A summary can be found in Wolff, 263–69, esp. 267–68 n. 23, who himself thinks that "doubtless the emperors did go a long way toward meeting the ethical demands of the new faith." Cf. the opposing view of J.B. Bury, *History of the Later Roman Empire* (London, 1889), 2:416: "The influence of Christianity on the legal conception of the conjugal relation was, as Zachariä remarks, small up to the time of Justinian; and it was the Isaurian Emperors who really introduced a christian legislation on the subject." Bury points out that it was only Leo, in 740, who instituted punishment for fornication and who forbade divorce, even by mutual consent, except on very narrow grounds (wife's adultery, husband's impotence, life-endangering slander by either, leprosy).
16. *LRE*, 974.
17. Cf. the chapter on "Conversion by Coercion" in R. MacMullen's *Christianizing the Roman Empire* (New Haven, 1984), 86–101.
18. C. Munier, *Concilia Africae,* Corp. Christ. 149 (Turnhout, 1974), 218: "In qua causa legem imperialem petendam promulgari."
19. See the cogent remarks of S. Mazzarino, *Aspetti sociali del quarto secolo,* Problemi e ricerche di storia antica 1 (Rome, 1951), 38–41.

(Mk 10:2-12; Mt 5:31-32, 19:3-12; Lk 16:18), where Jesus denounces divorce as Moses' concession to "hardness of heart" and not in accordance with God's plan for men and women. Paul on his own authority further counsels Christians not to divorce unbelieving spouses, urging them to hope for their conversion (1 Cor 7:2-16). Neither the Gospels nor Paul's letters, of course, are systematic treatises; and Paul does not give the church at Corinth any sanctions for disobedience to enforce on those who do not follow his prescriptions.

The church fathers down to the fifth century dealt with the subject of divorce on many occasions. Their reasons for forbidding it vary, as does their attitude toward marriage itself (some, for example, admitting affection as well as procreation as grounds for marriage), but they are mostly agreed that while divorce (in the sense of separation) is allowed for adultery by the other spouse, remarriage is not allowed to either party under any circumstance while the former spouse is still living.[20] The theologians and councils thus strongly urge the faithful not to take advantage of their civil right to remarry after divorce, even though many (a majority, in fact) of the theologians took as a command Jesus' permission to divorce an adulterous spouse (unclear though the interpretation of that passage is) and considered it mandatory. Ambrose (*Commentary on Luke,* 8.4) deals directly with the relationship of the church's teaching to civil law, in saying that human law allows divorce, but divine law forbids it. As a sacramental conception of marriage takes form with Augustine, a still stronger theoretical basis is provided for regarding marriage as indissoluble, at least at a spiritual level, while the partners lived.

In taking Jesus' radical statements on divorce at face value as ethical prescriptions for believers, the church had already from early days

20. For a careful and exhaustive treatment of the patristic discussions of divorce, see Henri Crouzel, *L'église primitive face au divorce du premier au cinquième siècle,* Théologie historique 13 (Paris, 1971), on which most of what follows relies. The subject is, of course, enormously controversial. Crouzel deals with many of his critics and those of a different persuasion in a collection of articles, *Mariage et divorce, célibat et caractère sacerdotaux dans l'église ancienne* (Turin, 1982). Ample bibliography will be found in both books. B. Löbmann, *Zweite Ehe und Ehescheidung bei den Griechen und Lateinern bis zum Ende des 5. Jahrhunderts* (Leipzig, 1980), though he lists (47 n. 59) Crouzel's book of 1971, seems hardly to have used him, repeating his examination in detail. He sees a more substantive theological division (eschatological vs. legalistic) between East and West; while he seems right that the West is more legalistic, it is not clear to me that this is the result of theological differences so much as of disciplinary and pastoral ones. The book is overall less satisfactory than Crouzel's. Much of the bibliography on divorce is apologetic and polemical in tone, seeking (especially since Vatican II) to support a particular view of how the Roman Catholic church of today should deal with divorced and remarried members.

taken a step it did not take with much of his other radical preaching.[21] Perhaps more important, affecting all areas of moral teaching, was the general development of a conception of church discipline which excluded from communion not only heretics but those regarded by the church as grievous sinners. This development did not take place all at once, but it was nearing maturity in the fourth century. It must be emphasized that this legalistic conception of regulating Christian behavior was equally a choice which could have gone differently. And even with that choice made, the formation of canon law was a long process, with no centralization or uniformity for a long time. It is anachronistic, I think, to look to the authors of this period for clear distinctions between behavior which is allowed and that which is only tolerated.[22] The authors tell us that certain behavior is contrary to divine law. A few of them, and some of the church councils, tell us that the church imposed penance or more severe penalties on certain behavior.

We find in the theologians and councils of the ancient church a considerable variety of opinion about how best to deal with the pastoral issues raised by the failure of many of the faithful to conform to the church's teaching on the issue of divorce. (None of them, however, except the council at Carthage cited above, sought to make the civil law conform to the church's teaching.) Not all of them took the view which ultimately prevailed in the West, that those who remarry after divorce should be excluded from communion. From the third and fourth centuries, the texts which pertain to the actual practice of pastoral clergy (as opposed to the abstract dictates of theologians) show some latitude. Origen cites the practice of some bishops in permitting remarriage to a woman while her husband is alive. He describes this behavior as contrary to scripture, but as intended to avoid greater evils,

21. See Crouzel, *L'église primitive,* 33, on Jesus' equally uncompromising teaching about oaths: "Or, non seulement l'Église a constamment admis le serment dans la vie publique et privée, mais elle a condamné à plusieurs reprises des hérétiques qui en refusaient la licéité et elle a imposé des serments aux membres du clergé, reconnaissant par le fait même qu'elle vit dans un monde où le "mal," c'est-à-dire le mensonge, existe encore, bien que l'idéal du Royaume comporte la sincérité absolute. Or elle refuse d'appliquer à la troisième antithèse le même réalisme, supposant ainsi que le baptême a radicalement supprimé la "dureté de coeur" de l'Ancien Testament, et elle impose à tous les séparés avec une rigidité juridique absolute, sous peine de refus des sacrements, la continence "en vue du royaume des cieux", comme si c'était "donné à tous" malgré les empêchements psychologiques et les insuffisances spirituelles."

22. On this point I believe that Crouzel is throughout his work too quick to use categories not really applicable at this date; that does not, however, diminish my agreement with his analysis of the authors' views.

and he thinks that the bishops have not acted unreasonably.[23] The Council of Elvira in Spain (probably 306), while maintaining on the whole a strict line, makes an exception in admitting to communion, in the event of infirmity, a woman who remarried after divorcing an adulterous husband.[24] The Council of Arles, in 314, ordered that husbands who put away unfaithful wives and were prohibited by the church from remarrying be advised (*consilium eis detur*) not to remarry while their wives lived; but no sanctions were imposed on them if they did remarry.[25] As these are the first two councils from which such canons are known, they are valuable evidence for the state of things in the early fourth-century West.

A more complex situation, which shows how much actual practice might vary, is found in the canonical work of Basil of Caesarea.[26] Basil's views are found in three letters of 314-315; all of them are based on and envisage specific situations. Basil allows hardly any reason for a woman to leave her husband; if she does, a woman who lives with the deserted husband is not condemned, he says. (The latter's behavior is classified as fornication but excused.) Basil says that the reason for the difference in customs for men and women is not easy to give, but that this is the way local custom goes. The view described by Basil (it is not at all clear that it is his own) is somewhere between the Roman and Pauline views of male/female equality in the matter of adultery, but closer to the Roman; the husband is treated as a fornicator with unmarried women, as an adulterer only with a married one. Even more striking is Basil's distinction in the matter of what behavior a spouse must put up with: a wife is to put up with almost anything, a husband not with adultery. Basil's practice and doctrine are thus different for the sexes, in that an abandoned husband can cohabit without being excluded from communion, whereas an abandoned wife cannot. Most striking of all is his admission that he cannot justify the local Cappadocian practice on the basis of the church's doctrine.

The apocryphal canons of Nicaea[27] seem to provide that a wife or husband unjustly accused by her or his spouse can repudiate him or her

23. Origen, *On Matthew* 14.23, cited by Crouzel, *L'église primitive,* 82–83, from *Die griechischen christlichen Schriftsteller* 10:340.25.
24. See Crouzel, *L'église primitive,* 116–21 on this text.
25. Crouzel, *L'église primitive,* 121–23; see C. Munier, *Concilia Galliae,* Corp. Christ. 148 (Turnhout, 1963), 11. Crouzel, *Mariage et divorce,* 88–91 and 127–42, discusses this canon, but his attention is focused on one textual crux, the outcome of which does not affect the question at stake here.
26. Crouzel, *L'église primitive,* 133–50.
27. Ibid., 240–43.

and remarry within the church. These canons come from an Egyptian source, perhaps as early as the start of the fifth century, but the date is not absolutely certain. They do not, however, give permission to divorce an adulterous spouse and remarry: if anyone does this, he or she will be chased from the church. There are also some Armenian canons[28] whose date is uncertain; some probably belong to 365. From them it seems that the Armenian church allowed remarriage after delay and penance.

The general conclusion to which these texts lead is that neither in the West nor particularly in the East was there uniformity in practice in the fourth and fifth century church in the discipline of communicants who remarried after a divorce. This does not mean that the church accepted and blessed such marriages, but it did in some instances show pastoral indulgence in dealing with the remarried member. We have no way of estimating how widespread exclusion was; the majority of the church councils do not deal with such marital issues at all.[29] Nowhere does the church invoke penalties other than public penance or exclusion from communion, nor suggest the application of the restrictions actually known to us from imperial laws, most of which concern property which the majority of communicants did not have.

On our second major question, the actual character of the late imperial legislation, two approaches are useful. First, the legal doctrine underlying the legislation. Wolff has shown that "the main pattern of thought which dominated Constantine's and his successors' divorce laws is already apparent in Augustus's well-known ban on the divorce of her husband-*patronus* by the *liberta*."[30] And after extensive study of the doctrinal foundations of the late imperial legislation, he concludes,

> divorce is forbidden, but once effected it cannot be brushed aside, since it always remained essentially the actual dissolution of a primarily social relationship. Only — in the *lex Iulia* — some of the normal consequences of a lawful divorce are eliminated, or — in the legislation of Constantine and his successors — the

28. Ibid., 244–46.
29. The Gallic councils hardly touch on the subject at all; the African ones only a little more. Ferrandus' *Breviarium Canonum,* drawn up in Carthage c. 523–546 on the basis of a wide variety of councils from West and East, includes only two rules (out of 232 canons) dealing with divorce (*Concilia Africae,* 287–306).
30. Wolff, "Trends," 279–88.

> unlawful divorce brings into play private and criminal law sanctions against him or her who disobeys the law.

In short, nothing in the late legislation either specifically adduces or implicitly rests on a doctrine of marriage different from the classical Roman one, so opposed in character to what we have seen of the church's understanding of the relationship.

Motives are harder. Wolff, though denying any doctrinal change in marriage law, thinks that "the Christian emperors were motivated by objectives entirely different from those of Augustus."[31] Is this true? The emperors are on occasions explicit about their reasons. First, as Theodosius II said in his *Novel* 12 (439), "the consideration of children demands that the dissolution of marriage must be more difficult." The remark is repeated at the start of *CJ* 5.17.8 (449). From the fact that all of the marital legislation is largely concerned with the disposition of dowry and marital gifts, we may surmise that what this means is that the consideration of the protection of the property of the children was at stake; the same is true of restrictions on remarriage.[32] The children of a first marriage were assumed to be injured by a second one. There was also a need for the safe-guarding of the separateness of the upper classes. We see both these motives at work, for example, in *CTh* 9.9.1, an edict of Constantine in 329: he prohibits the union of a women with her own slave. Both violators are to receive a capital sentence, and anyone has the right of prosecution. Any existing unions of this sort are to be broken up and the slave exiled; any children are to be deprived of the insignia of rank and their property. It is this last point which gives the game away, and it is brought home by later imperial legislation as well.[33] Most people did not have any insignia of rank to be taken from them, after all. Exactly the same motives lay behind Augustus' legislation on morals and marriage: he wanted senators to marry within their order, keep their marriages intact, be faithful, and produce lots of children. What the less exalted classes did, he could not have cared

31. Ibid., 287. He rejects, as too closely tied to the details, the detailed arguments of E.J. Jonkers, *Invloed van het Christendom op de romeinsche Wetgeving betreffende het Concubinaat en de Echtscheiding* (Wageningen, 1938), who emphasized the continuity of this legislation with earlier imperial enactments and identified in prechristian thought sufficient basis for the ideas which the emperors express. As we will see, however, when one tries to move from the specifics to the "general tendency," as Wolff calls it, one finds oneself in quicksand.
32. Cf. Jonkers, *Invloed*, 190, citing CJ 5.9.5.6.
33. See especially *NovAnth* 1, where the rule is extended to include freedmen as well as slaves, though existing unions are not broken up. Anthemius is explicit that it is the upper class which is the intended audience.

less,[34] and the same is true of his Christian successors more than three centuries later.[35]

These prudential considerations for family property and the integrity of the upper classes have nothing in common with the dogmatic basis of the Christian view of marriage described above, any more than the doctrinal basis underlying the actual edicts corresponds to a Christian understanding. And the laws themselves, which limit unilateral divorce to various acceptable grounds, allow divorce by mutual consent until Justinian, and permit remarriage after no lapse of time or one determined by the frivolity of the excuse for divorce, have no common ground with the church's view that remarriage after divorce was inconsonant with the nature of marriage in the first place. It seems, therefore, that neither doctrinal underpinnings, nor stated motives, nor actual results concord with a view that Christian influence is at work in the limits on repudiation.

And yet almost all writers on this subject have thought that the Constantinian and Justinianic limits on unilateral divorce proceeded from the influence of Christianity. For the most part, however, these writers have simply affirmed this influence, as if it were self-evident.[36] The few discussions to speak to the matter more explicitly are

34. Cf., e.g., R. Syme, *The Roman Revolution* (Oxford, 1939), 443–46.

35. The same attitudes can be seen in other areas of legislation. For example, social policy did not discourage sex for hire, as long as it involved women of the lower class. A law of Constantine from 326, classified in the Theodosian Code (*CTh* 9.7) as being on the law of adultery, makes this and much else clear: in cases of accusation, one is to check into the status of the woman. Tavern girls go free: "In consideration of the mean status of the woman who is brought to trial, the accusation shall be excluded and the men who are accused shall go free, since chastity is required only of those women who are held by the bonds of law, but those who because of their mean status in life are not deemed worthy of the consideration of the laws shall be immune from judicial severity." Serving women in drink shops were practically synonymous with prostitutes in antiquity, and it could hardly be clearer that (as three centuries earlier with Augustus) social class was the essential point of morals legislation. One sees no trace of any intrusion into this legislation of the uncompromising teachings of the gospels.

36. The literature is vast. As examples: C. Hohenlohe, *Einfluss des Christentums auf das Corpus juris civilis* (Vienna, 1937), an uncritical and pious account; J. Gaudemet, *La formation du droit séculier et du droit de l'église aux IVe et Ve siècles,* 2 ed. (Paris, 1979), who describes Christian influence on family legislation as "indéniable, encore que plus limitée qu'on ne l'a cru parfois"; J. Vogt, "Zur Frage des christlichen Einflusses auf die Gesetzgebung Konstantins des Grossen," *Festschrift für Leopold Wenger,* Münchener Beiträge 35 (Munich, 1945), 2:118–48 at 134: despite the lack of resemblance of Christian and legal views, he insists on a connection (admitting, however, that the punishment

unsatisfactory, no doubt because motives which find no expression and which cannot be deduced from the doctrine of marriage at work in the laws are hard to demonstrate.[37] For example, Wolff cites in support of "the general tendency which, especially with Constantine and again with Justinian...was directed toward strengthening the marital bond in the Christian sense" a work of Biondo Biondi on Justinian, in which Biondi proclaims that Justinian was "the first Christian emperor who promulgated a truly Christian law," referring to *Novel* 117.10. "Riforma audace che di tanto si allontanava dalla tradizionale concezione pagana del matrimonio di quanto rispondeva alla concezione cristiana," says Biondi.[38] Now the Novel in question does not offer any justification for its denial of consensual divorce, any more than for other provisions. Insofar as one sees a consistent theme throught *Novel* 117, it is (as usual) the protection of the property rights of the children of repudiated marriages. The one overtly "Christian" element is the exception made for those who wish to devote themselves to a chaste religious life[39]; but such separation is in general not approved of by the main Christian theologians of the period before Justinian.[40] Biondi's affirmation is in fact nothing more than an act

of a second marriage is only a matter of protecting the children of the first, not of a Christian view of remarriage); cf. Wolff, "Trends," 267 n. 23 for further bibliography, and a more detailed report on the literature concerning Christian influence on late Roman law by L. Wenger in *Archiv für Papyrusforschung* 12 (1937): 298–307.

37. Cf. V. Basanoff, "Les sources chrétiennes de la loi de Constantin sur le repudium," *Studi in onore di Salvatore Riccobono* (Palermo, 1932), 3:177–99, who bizarrely describes the penalties in the Constantinian law as (194), "la formule de transaction inspirée, selon toute vraisemblance, par le texte du Pasteur, adapté à des conditions nouvelles." It is hard to see what possible connection between Hermas' teaching and the Constantinian legislation can have provoked this remark. Equally odd is his insistence that classical Roman law knew no divorce by consent (193); since the decision of one person was sufficient, surely that of two was also adequate, and classical law required no particular form of divorce.

38. *Giustiniano primo, principe e legislatore cattolico,* Pubbl. Univ. Catt. del Sacro Cuore, 2 ser., Scienze giuridiche, 48 (Milan, 1936), 43.

39. *Castitatis concupiscentia,* in the deliciously oxymoronic phrase of the apparently contemporary Latin version of the law preserved in the Appendix to the Epitome of Iulianus (*sophrosynes epithymia* in the original Greek); Biondi apparently did not appreciate the phrase, for he substitutes *desiderium* (from the translation of the editor, Schoell) for *concupiscentia* in quoting it! One should also note relegation to a monastery as punishment for contravention of these laws by women — hardly a Christian view of the monastic vocation.

40. See Crouzel, *L'église primitive,* 376–77. Such separation is forbidden by a canon from a Council of Gangra cited in Ferrandus' *Breviarium*, in *Concilia Africae*, 301, no. 164: "Ut si qua mulier quasi religionis causa virum dimiserit, anathema sit."

of faith in his "catholic" Justinian.[41] Now one cannot demonstrate Justinian's state of mind before ordering this legislation, and that a Christian emperor wanted to discourage divorce may be quite right; but to affirm his motives, in the total absence of evidence, is the work of a hagiographer, not an historian.[42]

We now turn to inquire what the effect in practice of all of this legislation was. An enumeration and description of the surviving documentation for divorce is enlightening. Traditionally, Egyptian law had, like Roman law, allowed divorce either by mutual consent or by repudiation, with no penalties for either party unless some had been established in a contract concerning property.[43] In Roman Egypt up to the fourth century, these compatible traditions produced a consistent habit of divorce agreements by mutual consent, in which both parties are freed of liability.[44] The first of our fourth-century documents (*P.Oxy.* XXXVI 2770, of 26 January 304) shows the expected pattern: Herakles and Maria[45] agree that they are divorced, that each has his own property, and that they have no claims against each other. There are no children, each is able to remarry at will. The document itself is called τὰ τῆς 'αποξυγῆς γράμματα, the document of divorce. A similar formula is found in the undated *P.Oxy.* XLIII 3139 (late third/early fourth century), which calls itself τὰ τῆς περιλύσεως καὶ 'ανεγκλησίας γράμματα, the document of dissolution (of marriage) and of renunciation of claim. From the following year comes *M.Chr.* 295 (= *P.Grenf.* II 76 = *Jur.Pap.* 21, 305/6, Great Oasis). It begins "Since, as a result of some

41. "Certo non tutta l'attività di Giustiniano è encomiabile," writes Biondi, *Giustiniano,* 189; but he demurs only at his excess of zeal, and the purpose of his book is overtly encomiastic, a hymn of praise to Justinian for supposedly making civil law conform to the law of the church.
42. It should, however, be mentioned that the quaestor responsible for issuing the legislation was evidently Tribonian's successor Junilus, who seems to have been a (lay?) theologian as well; cf. A.M. Honoré, *Tribonian* (Ithaca, 1978), 237–40.
43. See P. W. Pestman, *Marriage and Matrimonial Property in Ancient Egypt,* Pap. Lugd.Bat. 9 (Leiden, 1961) for a comprehensive treatment; there is also an interesting article by S. Allam, "Quelques aspects du mariage dans l'Egypte ancienne," *Journal of Egyptian Archeology* 67 (1981): 116–35, stressing practical deterrents to behavior as free as theory would allow.
44. The standard list is O. Montevecchi, *Aegyptus* 16 (1936): 20, with additions in her *La papirologia* (Turin, 1973), 206. A comprehensive treatment of divorce in the later period can be found in Andreas Merklein, "Das Ehescheidungsrecht nach den Papyri der byzantinischen Zeit" (Dissertation, Erlangen-Nürnberg, 1967).
45. The editor remarks, "on Tcherikover's criteria…she should be considered as Jewish. But since her family and husband's family bear Graeco-Egyptian names, and the document itself offers no other indication of her religion, it is possible to accept her as a pagan, believing her name to be a reflection of Jewish or Christian influence." In a fourth-century context, surely a Christian identity is more likely.

evil demon's having come upon us, we have agreed to be divorced from one another with respect to our common marriage," and again proceeds to declare that neither party has any obligations to the other. All of these thus have clear antecedents in the Egyptian practices of earlier centuries; all (probably) are prior to Constantine's legislation.[46]

Regrettably, we have no actual divorce documents from the thirty-year period after Constantine's legislation limiting repudiation. The next such document is found only with *P.Stras.* III 142, of the year 391. The core of the contract, in the form of a letter from the woman to the man, runs as follows:

> Since I, Allous, lived with you, Elias, for some time, but we decided for some reason of an evil demon which came upon our common life together to separate, in accordance with this I, Allous, agree that I have no claim against you concerning our life together or any other written or unwritten debt or collection or claim or inquiry at all, and that you Elias have the right to cohabit in another marriage, with you being free of complaint about this.

Blame is missing: some evil demon came upon the relationship, it ended, both parties are free, and no one has any claim on the other. The document describes itself as a *perilusis*. It is obviously close in character to earlier documents of the type. A similar document seems to be referred to in *P.Lips.* 39 (390), a petition in which a woman states that she sent her former husband a *repudium* after a *dialysis*, a mutual dissolution of the marriage, executed between them. (Her current complaint concerns his subsequent assault on her.)

The terminology here is of great interest. *Repudium* is linked to *dialysis*; evidently an agreement to divorce was followed by at least the wife's, and probably both parties', sending to the other a document stating that they were divorced and that they had no claim against the other (though such documents were evidently not required until 439).

46. The one new element in this century is the "evil demon" phrase. This is treated in detail by Merklein, *Das Ehescheidungsrecht,* 73–79; he concludes that it is a means of avoiding any question of the culpability of one spouse or the other which might lead to later legal problems. R.C. McCail, *Mnemosyne* 21 (1968): 76–78, points out that the evil demon appears in Justin's *Novel* 140 ("Therefore we pray that marriages may be happy for those who enter on them, so that they never become the work of the evil demon") and argues for a reference to its *phthonos* in *Anth.Pal.* 7.596 (Agathias). These passages suggest that it is not used merely as a legal self-exculpation in the papyri, but was a widespread conception.

This linkage, observed here first in 390, continues in later documents. Regrettably, we have no actual divorce documents from the fifth century, but sixth-century terminology is consistent with the Leipzig papyrus. *P.Herm.* 29 (586), executed between two Samaritans, presents itself is ἀντισυγγραφαὶ, mutual agreements, of διάλυσις ἤτοι 'ρεπούδιον; each party is to have a copy for security. The formula is familiar:

> Whereas we were of late joined together in lawful marriage and community of life and for the procreation of children according to the usage of men, with good hopes; and today, we know not whence, through some malign spirit, they decided to part from each other....

Claims are renounced. Another mutual agreement of divorce from 569 (*P.Flor.* I 93 = *M.Chr.* 297) describes the document as a *repudium*. The term could, of course, be used also of a unilateral declaration, like that by which the father of a married woman who was still in his *potestas* terminated her marriage on the grounds of "lawless deeds, which are pleasing neither to God nor man and are not fit to be put into writing" on the part of his "most honorable son-in-law" (*P.Oxy.* = *M.Chr.* 296; translation from *Select Papyri* I 9). This latter declaration is addressed to the son-in-law through the *defensor civitatis (ekdikos)*.[47]

What we find so far, then, is substantial continuity, in substance and phraseology, from the time before Constantine to the late sixth century, after numerous changes in imperial laws. It is true that we have no divorce documents from the period between Constantine and Julian, nor from the fifth century, although *P.Cair.Masp.* II 67154 from the reign of Justinian (which year, we cannot tell[48]) does provide an example of divorce by mutual consent in the first half of the sixth century. It is sworn to by the Holy Trinity and the emperor, thus breaking impartially Jesus' bans on divorce and oaths in one breath. But the fifth century is not in any case a well-documented period. A

47. Other sixth-century references to *repudia* are *P.Cair.Masp.* I 67121 (573), II 67153 (=67311) (568), 67154 (Justinian), 67155 (date lost). Several of these contain the same language as the earlier divorce agreements, blaming an evil demon; all are mutual agreements. Merklein, *Das Ehescheidungsrecht,* 64–67, thinks that the papyrological use of *repudium* does not indicate a full "reception" of the Roman concept in Egypt.

48. The date is provided only by an oath formula invoking Justinian; the editor thought that the text on the other side of the papyrus was written "assez longtemps" after the recto, and specifically in the period 566–570. But that does not help us to know whether the divorce fell before or after 542.

review of our scanty material on marital relations from the period will bring us back to the document with which we started.

In a papyrus of 362,[49] a petitioner alleges that six years earlier he married a woman named Tamounis and gave her marriage gifts. He carried out all of the legal and normal duties and lived with her three years. But her mother took her away and gave her to another man in marriage while the complainant was away on private business trying to make enough to live on. When he came back... the papyrus breaks off. After three years, what redress is he seeking? We cannot tell. But it is worth pointing out that three years before was in 359, before any possible Julianic abrogation of Constantine's legislation. (Indeed, the date of that abrogation seems to be 363, as we have seen, or after the date of this papyrus.) There is no question of mutual consent to divorce in this case[50]; of course, since we do not have the mother's side, we do not know what cause she can have alleged. Since Constantine left wives hardly any usable causes for repudiation, however, the mother's action would on the face of it appear to violate the law. After three years, no consequences have followed.[51]

That we have no traces of the effects of any of the imperial legislation on divorce should not unduly surprise us.[52] Modern studies of such areas as civil procedure and restrictions on social mobility have concluded that the effectiveness of imperial legislation in later Roman Egypt was very limited. Schiller has shown that there is no evidence for civil procedure in sixth-century Egypt, a fact he suggests may in part be a result of Coptic disaffection with the Chalcedonian central government.[53] In fact, Schiller argued that in the sixth century "contemporaneous legislation emanating from Constantinople likewise had no impact on the current law of Egypt."[54] On the other side, Taubenschlag gave a very optimistic account of the effectiveness of

49. *P.Cair.Preis.* 2 (Hermopolis; cf. BL 3.36).
50. *P.Stras.* III 131, a marriage agreement of 363, specifically foresees divorce by mutual consent if differences arise between the couple.
51. Parental meddling turns up in another petition of earlier date, and was no doubt a common phenomenon: *P.Sakaon* 38 (Theadelphia, 312).
52. Merklein, *Das Ehescheidungsrecht,* 102–4, notes that there is no sign of the enforcement of imperial laws on divorce in Egypt, nor of any other curtailment of freedom of divorce.
53. A.A. Schiller, "The Courts are no More," *Studi in onore di Edoardo Volterra* (Milan, 1969), 1:469–502, esp. 498–502.
54. A.A. Schiller, "The Fate of Imperial Legislation in Late Byzantine Egypt," *Legal Thought in the United States of America under Contemporary Pressures,* ed. J.N. Hazard and W.J. Wagner (Brussels, 1970), 41–60 at 45.

Justinian's legislation on divorce,[55] but without any real basis, for his only item of interest is the father's repudiation of his daughter's husband in *P.Oxy.* I 129, which would have been equally possible at any time before Justinian and thus need not reflect his legislation at all. The fourth and fifth-century situation is less clear. Keenan showed that late antique Egypt's population seems neither more nor less mobile than that of earlier periods, despite imperial legislation; he cautions, however, against excessively sweeping conclusions about the effectiveness of imperial laws.[56] Schiller, who thought that fourth and fifth-century laws had some effect in Egypt through the efforts of private notaries (rather than through direct governmental enforcement), placed the end of that effectiveness in the early sixth century.[57] In any case, imperial legislation directly addressed to officials in Egypt in almost all cases concerns matters of public law (liturgies, taxes, and the like), not the civil law.[58]

An undated fourth-century affidavit (*P.Oxy.* VI 903) provides an interesting glimpse which leads us back to the violent marital conflicts of *P.Oxy.* 3581 and the church's involvement in these marriages. The couple in question lived together without any written documents; both were evidently well-off slaveowners. There were disputes over property, with the man suspecting theft by the staff.

> He shut up his own slaves and mine with my foster-daughters and his overseer and son for seven whole days in his cellars, having insulted his slaves and my slave Zoe and half killed them with blows, and he applied fire to my foster-daughters, having stripped them quite naked, which is contrary to the laws. [Other insults are recounted. Then there was an agreement, in the presence of the bishops.]
>
> "Henceforward I [the husband] will not hide all my keys from her (he trusted his slaves but he would not trust me); I will stop and not insult her." Whereupon a marriage contract was drawn up, and after these agreements and the oaths he again hid his keys.

55. "The Legislation of Justinian in the Light of the Papyri," *Byzantion* 15 (1940–41): 280–95 = *Opera Minora* (Warsaw, 1959), 2:69–89 at 73–74.
56. J.G. Keenan, "On Law and Society in Late Roman Egypt," *Zeitschrift für Papyrologie und Epigraphik* 17 (1975): 237–50. He gives extensive bibliography on earlier discussion of the general topic of the effectiveness of imperial laws in Egypt.
57. Schiller, "The Fate," 54–58.
58. See V. Dautzenberg, "Die Gesetze des Codex Theodosianus und des Codex Justinianus für Aegypten im Spiegel der Papyri" (Dissertation, Cologne, 1971).

She went to church on the sabbath, and he asked her why. He did not pay the 100 artabas of wheat due on her land and prevented her from doing so. His assistant was taken to prison. And so on. He ordered her to send away her slave Anilla. He said, "After a month I will take a courtesan for myself." "God knows these things," concludes the woman.

The role of the bishops (perhaps of Oxyrhynchos and Antinoopolis) here is as reconcilers and witnesses or guarantors of the reconciliation. This is, of course, the same role that the presbyters, priests, exercised in *P.Oxy.* 3581, although in retrospect it looked like being "deceived" by the man through the priests. There is not in either document any indication of any disciplinary attempts or coercion by the clergy; rather, so far as we can see, these texts are evidence for a reconciling pastoral role. We may perhaps have evidence of a similar sort of work in *P.Grenf.* II 73, where we find one priest sending someone who is probably a prostitute to another priest, who finds people to take care of her until her son arrives.

There is no trace in the papyri of any sentiments depreciating the body, sexual expression, or marriage, despite the growth of monasticism during this very century. There are, to be sure, signs of that monasticism and its possible difficulties with civil society: mention of a consecrated virgin in conflict with what I suppose to be some relatives over ownership of some Christian books,[59] and an interesting letter dated by its handwriting to the fifth century.[60] Written to someone who is evidently the head of a religious community on behalf of a third party, probably described as a priest, it contains two key sentences:

> Let me not then fail of my petition; for he took his wife [*eleuthera*] a long time ago, and, in my view, cannot divorce her. So I beseech you to have compassion for his unhappy plight and to give him a decision to enter his topos [i.e., religious community].

That marriage was excluded from the life of those entering monastic communities can hardly surprise us; what is interesting is the appeal to let a married man enter the community without divorcing his wife.[61]

59. *P.Lips.* 43 (Hermopolite).
60. *P.Herm.* 16 (5th century).
61. The fathers, following Paul, generally refused to countenance separation of a married couple so that one could enter monastic life; cf. Crouzel, *L'église primitive,* 376–77.

Though the maintenance of the marriage may for all we know have been of more significance for property than for any personal relationship, the writer's stance is marked by concern and compassion for the personal situation rather than legalism.

When Christianity spread and became publicly visible in fourth-century Egypt, it faced a country with traditions which favored easy marriage and easy divorce, and in which the consent of both parties to both was the principal ingredient. There is no indication that in practice it tried to enforce its view of marriage by widespread ecclesiastical discipline[62] of the sort which it had developed earlier when Christians were a small, cohesive minority.[63] The eastern churches, in medieval and modern times, have, it is true, allowed divorce followed by remarriage in many cases where the western church did not, a position often viewed in the West as "laxity."[64] The Coptic church, for example, has historically allowed divorce for adultery, for the nonvirginity of a bride, for entry into monastic life, and in certain other cases.[65] But the situation we see in the papyri does not show us the church legalistically enforcing the continuance of marriage subject to certain escape routes. What we do find it doing, rather, is acting to encourage marriage, to reconcile spouses who were at odds, to sustain marriage, even to help those on the edge of social

62. An interesting comparison is afforded by *P.Ness.* 57, a divorce agreement of 689 from Nessana in Palestine, in which a priest and his wife part company before seven witnesses who include a priest, an archdeacon, and a deacon. He retains the dowry without contest; she seeks only her freedom ("We want nothing from you.... Only release me.")

63. Those divorcing were certainly mostly Christians. As Merklein, *Das Ehescheidungsrecht,* 75, notes, many late divorce documents have explicit marks of Christianity, such as crosses at the top and other such elements. It should be added that most divorce documents include explicit guarantees of the ability of the other party to remarry at will. I rather suspect that in this period visible disciplining of members for divorce or remarriage was limited to prominent and wealthy persons whose example might be hoped to do some good.

64. Cf. Ignaz Fahrner, *Geschichte der Ehescheidung im kanonischen Recht* 1: *Geschichte des Unauflöslichkeitsprinzips und der volkommenen Scheidung der Ehe im kanonischen Recht* (Freiburg i. Br., 1903), 33–34. The medieval eastern churches had a much closer relationship with the government than the western church, with mutual influence: the emperors required church solemnization of marriage, while the church tended to accept the government's definition of grounds for divorce. For the latter phenomenon, see J. Dauvillier and C. DeClercq, *Le mariage en droit canonique oriental* (Paris, 1936), 85.

65. See Luigi Bressan, *Il divorzio nelle chiese orientali* (Bologna, 1976), 36. Bressan is concerned mostly with the Orthodox churches, which have a complicated history of allowing divorce on grounds of adultery as well as other bases.

acceptability as prostitutes.[66] Just as in the Cappadocia of Basil,[67] this picture reflects the reality of a pastoral church with limited power over its members[68] at work in a society with a well-defined set of traditional standards for relationships between men and women and a normal assortment of the standard human failures that mark such relationships.[69]

■

66. And we find at least more of the rhetoric of family affection in private letters in Christian circles than otherwise. Cf., e.g., *P.Ant.* II 93.

67. Though with quite a different (and much more egalitarian) set of cultural standards for behavior than the Cappadocians'.

68. A factor not limited to this time and place. Cf. P. Ariès, "The Indissoluble Marriage," *Western Sexuality: Practice and Precept in Past and Present Times,* ed. P. Ariès and A. Béjin (Oxford, 1985), 140–57, who argues that it was not until the twelfth century in France that the church concentrated its efforts, with some success, on the indissolubility of marriage.

69. The Orthodox churches developed in time a doctrine of divorce as being a matter of *oikonomia,* or pastoral understanding and ministry to those who fail to live up to its norms. Bressan, *Il divorzio,* 43–44, regards this view as a recent and (as explicit doctrine) minority view, but one with unstated roots in long-standing pastoral practice. It is set out, with a clear rejection of a legalistic approach to the biblical norm, from a contemporary perspective in J. Meyendorff, *Marriage: An Orthodox Perspective,* 2nd ed. (Crestwood, NY, 1975), 60–65. Meyendorff's modern approach resembles strikingly what we can see of actual pastoral practice in the ancient Egyptian church.

IV
Literary Criticism in the Platonic *Scholia*

JAMES COULTER

THE STANDARD COLLECTION of the *scholia* to Plato's dialogues, edited by William Chase Green, contains marginal annotations deriving from five witnesses to the text of Plato's dialogues — B, W, A, T, and C.[1] For those not familiar with Greene's edition or with the chief conclusions of scholarship devoted to the *scholia* some introductory remarks may be helpful. First, along with the so-called *scholia vetera* found in the codices, that is, those *scholia* which are the residuum of ancient scholarship, there are preserved in the Oxford codex (B, which comes from the library of Archbishop Arethas of Caesarea) a large number of marginal annotations in Arethas' own hand, some drawing on ancient commentary, but most of his own composition. These Arethan *scholia* Greene prints in a separate section at the back of his edition.[2] Although this is a justifiable procedure for the most part, those *scholia* of Arethas which are in character true *scholia vetera* are as a consequence condemned, somewhat illogically, to a second-class status. Secondly, it needs to be kept in mind that by limiting the *scholia* which he has admitted into his edition to those from the five major witnesses, Greene denies the reader access to some interesting annotations found in other codices. These the reader must still consult in Hermann's 1853 Teubner edition, where however as an unhappy

1. See *Scholia Platonica*, ed. William Chase Greene, American Philological Association. Philological Monographs 8 (1938). B = Oxford, Bodleian Library, MS Clarkianus 39, written in 895. W = Vienna, Österreichische Nationalbibliothek, MS Suppl. Phil. Gr. 7; the portion containing relevant *scholia* was probably written in the eleventh or twelfth century. A = Paris, Bibliothèque Nationale, MS Graecus 1807, written in the early or middle ninth century. T = Venice, Biblioteca Nazionale Marciana, MS Append. Class. 4.1, the relevant portion written in the late eleventh or early twelfth century. O = Vatican City, Bibliotheca Apostolica Vaticana, MS Vaticanus Graecus 1, written in the tenth century. There is a valuable review of Greene's collection by W.A. Oldfather in *Classical Philology* 36 (1941): 371–89. The *scholia* to the *Gorgias* have recently been published separately: Mirella Carbonara Nadei, *Gli scoli greci al Gorgia di Platone: Testo, traduzione e note* (Bologna, 1976).
2. Pages 417–80.

consequence of Hermann's editorial practice the reader is not always able to determine the exact provenance of a given *scholium*.

The sources of these older *scholia* were intensively investigated in the period from about 1880 to 1920 by, among other scholars, Wolf, Cohn, Mettauer, Alline and Gudemann, and the results of this work are usefully summarized in the preface to Greene's edition.[3] These studies show that one important group of sources for the *scholia* are the ancient *lexica*, not only those of the Atticists but also such special works as Timaeus' Lexicon to Plato. In another group of *scholia*, which do not draw upon the *lexica*, there are discussions of matters historical, geographical, mythological, and paroemiographic, some based on sources still extant, such as Strabo, Ptolemy, and Hesychius, others from scholarly works which have not survived save as traces in later works which made common use of them with the compilers of the Platonic *scholia*. From our perspective, however, the most important source by far of the *scholia* are the extensive commentaries on Plato's dialogues composed in the Neoplatonic schools of Athens and Alexandria in the period from about 450–575 C.E. Of these commentaries eleven are extant, not all, however, in their original or complete form. Embodying a long tradition of school commentary they expound the following nine dialogues: *Phaedo* (two commentaries), *Cratylus, Parmenides, Philebus, Phaedrus, Alcibiades* I (two commentaries), *Georgias, Republic, Timaeus.* The commentators are Proclus of Athens (410–485), Hereias of Alexandria (fl. c. 450), Damascius of Athens (c.485–after 533), and Olympiodorus of Alexandria (fl. second half of the sixth century).[4]

A GREAT NUMBER of the Platonic *scholia* and, more pertinently, virtually every *scholium* which has any interest as literary-critical comment are taken either verbatim or in paraphrase from these commentaries — a fact which constitutes a rare circumstance in the history of annotation. There are other cases, to be sure, where both *scholia* and ancient commentaries which lie behind them are still extant and available for

3. Pages xxv–xxxiii.

4. There seems to be no adequate survey of the Neoplatonic commentary literature available in one place, but the introductions to the following works, taken together, give a valuable and authoritative picture of the commentators and their setting: Proclus, *Théologie Platonicienne,* vol. 1, ed. H.D. Saffrey and L.G. Westerink (Paris, 1968), esp. ix–lx; *The Greek Commentaries on Plato's* Phaedo, vol. 1, *Olympiodorus*, ed. L.G. Westerink (Amsterdam, 1976), esp. 7–29; idem, vol. 2, *Damascius*, ed. L.G. Westerink (Amsterdam, 1977), 7–18; *Anonymous Prolegomena to Platonic Philosophy,* ed. L.G. Westerink (Amsterdam, 1962), esp. x–xxv.

comparison: Aristotle is one instance. There, however, the texts commented on are technical works. What is true of Plato is true of no other author who was also a great literary artist. It would be as if we still possessed along with the *scholia* on Aristophanes' plays the complete commentaries of Symmachus and Didymus or, in the case of Homer, the full text of the works which form the basis of the so-called Viermännerkommentar of the *scholia.* With regard to the Platonic *scholia*, however, the fact that those deriving from the Neoplatonic commentaries should have turned out to be the only ones to present any interest from the perspective of literary criticism is not surprising when one considers the exegetical bent of the Neoplatonists. I have, in fact, tried to show elsewhere how powerful a system of textual interpretation of Plato's dialogues can be extracted from their commentaries.[5]

THERE IS A FURTHER POINT worth considering, which is somewhat disquieting in its implication. Let us suppose that in the case of Plato, as in that of the other great classical writers, we did *not* have the ancient commentaries as well as the *scholia* derived from them. What theoretical sense *could* we make of the brief, often elliptical and fragmentary observations one regularly finds in the *scholia*? In the Neoplatonic commentaries themselves remarks on individual passages in the dialogues are always, in principle at least, specific applications of general canons of interpretation which the commentators themselves explicitly articulate, often at great length. The question that will be addressed in this paper, then, is whether, if we did not happen also to possess the commentaries, we should be in a position to reconstitute the conceptual system in terms of which alone the real significance of these separate observations is to be understood. To try to answer this question, let us turn to some of the more notable *scholia*, notable, that is, for the attempts at literary criticism which they contain.

A useful group with which to start is that in which we perceive an effort to define, in the light of what is to be presumed to be Plato's compositional intention, the precise significance of some detail in the narrative. Thus, at the very beginning of the *Phaedrus*, Socrates, in reply to a question concerning Lysias' whereabouts, is told by Phaedrus that the orator is staying with Epicrates in a house belonging to a certain Morychus

5. See James A. Coulter, *The Literary Microcosm: Theories of Interpretation of the Later Neoplatonists,* Columbia Studies in the Classical Tradition 2 (Leiden, 1976).

(227b). In a *scholium* taken from Hermeias' commentary we read as follows[6]:

> Morychus was glutton, attacked also by the comic writers for his gluttony. It is therefore reasonable (*eikos*) that the intemperate (*akolaston*) Lysias should be staying at the house of such a man.[7]

In light of Plato's analysis of non-philosophical rhetoric later in the *Phaedrus* and of his hints about the real nature of the intentions behind the words of the non-lover of Lysias' speech, this observation, arbitrary as it may seem, is not without genuine insight. As Plato had already argued in the *Gorgias*, the rhetoric of his time should be thought of as an activity no more beneficial to the soul than gourmet cooking was to the body. Immediately after the passage the *scholium* comments upon, in fact, Socrates asks a further question in response to Lysias' answer: "Well, then, how did you spend your time there? Lysias, I suspect, entertained you with a *banquet* of discourse."

From the *Phaedo* comes another example illustrating precisely the same approach to evaluating narrative in Plato. It is an observation on Plato's description of the way Socrates changes his position on his couch (*kline*) just after he has urged Evenus, and anyone else who is a true philosopher, to follow him in death as soon as he could (*Phaedo* 61c-d). By way of caution Socrates adds that one must, of course, not allow violence on oneself to attain this philosophically desirable state. Suicide is taboo (*ou themiton*). "As he was saying this he put his legs on the ground, and it was sitting in this position that he carried on the rest of the conversation." In this conversation the first question is Cebes', and it is about Socrates' prohibition against suicide. For most readers Plato's description is moving in its homely vividness. The reasons for this are not easy to formulate, but they do seem to reside to some degree on the impression Plato creates, through a precise description of concrete detail, of a powerful summoning up of intellectual energies on Socrates' part on this last day of his life. The *scholium*, in fact, conveys a very similar sense of the effect of Plato's description:

6. P. Couvreur, *Hermiae Alexandrini in Platonis Phaedrum scholia*, Bibliothèque de l'École des Hautes Études, Sect. des sciences historiques et philologiques 133 (Paris, 1901), 18, 12.
7. Page 68.

> Socrates has assumed a more earnest and majestic posture (*syntonôtero... schêma kai semnoteron*), because it is his purpose to converse on a more momentous subject (*peri problematos semnoterou*).[8]

So, too, the portrayal of Socrates as compelled to reformulate for Callicles the same question over and over again conveys a powerful sense of the latter's reluctance (*oknos*) to concede the point which will signal his defeat (*Gorgias* 489a).[9]

In these instances — and there are others — the assumption underlying the comments in the *scholia* is that the interpreter dealing with Platonic narrative should press the details which make it up for all their significance.[10] He will do this in the confident belief that Plato's most serious intentions will be discerned in even the most apparently trivial particulars. According to this view, the narrative details of the dialogues are there to convey Plato's deepest philosophical concerns and never merely to charm the reader or to build up an illusionistic sense of verisimilitude. As we shall see when we outline the views of the Neoplatonic commentators who are the source of these *scholia*, *all* narrative detail in the dialogues must be understood as expressions of a single conscious, guiding intellect. This belief should, of course, come as no surprise. The Neoplatonists were, after all, philosophers, expounding the texts of another philosopher who, as we may recall, urged serious second thoughts upon anyone misguided enough to suppose that written works could be composed without their author first acquiring a grasp of the truth which is accessible to the accomplished dialectician alone.

Another group of *scholia*, more rhetorical in orientation, is concerned with what characters *say*. Their concern is with defining the stylistic effect or the social, typical nuances of a characters words, and they presuppose a desire on Plato's part to have his interlocutors speak "in character." Wishing, for instance, to underscore the fact that Gorgias' fervent disciple Polus speaks just like his master (*Gorgias* 448c), the *scholium* gives us the following nudge: "Look at Polus' balanced clauses (*ta parisa*).[11] Similarly, we are invited to note in our reading of *Alcibiades I* the way in which Alcibiades words evince both

8. Greene, 8.
9. See also Greene, 152.
10. Compare the *scholia* to *Phaedo* 61e (Socrates must wait until sunset to be executed), and *Gorgias* 523e (why the judge in the underworld is naked).
11. Greene, 131.

his intellectual slackness and his youthful pride and arrogance.[12] At *Laws* 626c Megillus the Spartan endorses Cleinias' assurance to the Athenian Stranger (falsely identified in the *scholium* as Socrates)[13] that he, Megillus, thinks as Cleinias does; he says, "What other answers, O godlike man, would any Spartan give?" The *scholium* reads as follows

> His reply displays the Spartan manner. It is brief (*syntomos*), compressed (*synestrammenê*) and suggests at the same time a ruggedness of style (*kata tên lexin trachytêta*).[14]

These *scholia* are concerned with what may be called the outer shell of character — the broad or obvious signs of social or generic identity. In contrast, there are times when the commentators believed that Plato has his characters use words which are intended to convey a private emotion or a particular, momentary state of mind. The *Phaedo* provides another example, this too from the passage on the correct philosophical attitude toward suicide. When Socrates remarks that it may seem strange, even in the case of a man for whom death is preferable to life, that he should await a "benefactor" who will do him this kindness, Cebes, "laughing gently in response to Socrates' words, said in his native dialect, 'May Zeus be my witness (*Ittô Zeus*)!'" (*Phaedo* 62a). The interpretation of the *scholium*, which goes back to at least Olympiodorus,[15] is that this oath is to be understood as expressing, specifically, Cebes' amazement at Socrates' use in this somber connection of the word "benefactor." The *scholium* continues:

> *Ittô* is the form employed by a native of Boeotia, and Cebes' use of it shows that he is genuinely astonished at Socrates' (*physei ton Sokratên thaumazei*).[16]

The assumption which underlies this observation is, it seems, that in moments of strong or deep emotion we revert to the earliest, more spontaneous layers of our personality. It is, I think, a psychologically plausible observation.

12. Respectively, *scholia* to *Alcibiades I* 130c (Greene, 105) and 112d (Greene, 451).
13. See the discussion of L. Tarán in his *Academica: Plato, Philip of Opus, and the Pseudo-Platonic Epinomis,* Memoirs of the American Philosophical Society 107 (Philadelphia, 1975), 131, n. 548.
14. Greene, 298.
15. Westerink, *Olympiodorus* 1:59.
16. Greene, 10.

In addition to amazement, states of mind such as ambivalence, resistance to argument, aggressiveness, and, of course, irony are discerned by the commentators. Most often, to be sure, the experienced reader will find these indications annoyingly obvious. Thus, the "figure" (*schêma*) Callicles employs in addressing Socrates — referring to him in the third person as "this fellow here" — is a sign of his "audacity" (*thrasytês*).[17] Unfortunately, in regard to the interesting matter of Socratic irony the *scholia* almost always point out only the blatantly obvious.[18] Containing some small element of interest, however, are the following two observations: the use of the particle *dê* to reinforce an ironic nuance[19] and the notion that speaking *kat' antiphrasin* (*per oppositionem*) is a closely related variant of irony.[20]

There is another group of *scholia* which, rather than containing surmises concerning the speaker's feelings or states of mind, argue instead for some conscious intention lying behind the speaker's words. One instance is the sarcasm of Callicles' words mimicking Socrates' intentionally formal manner of addressing him in the full form, that is, by his patronymic and demotic.[21] At the point in the *Theaetetus* (162d) also, where Socrates in conversation with Theodorus has just concluded a playful refutation of Protagoras' claim that man is the measure of all things, Theodorus urges Socrates to invite the young Theaetetus to take his place, and Socrates agrees to do so. Theaetetus is understandably reluctant, and Socrates, exhibiting the sensitivity to his interlocutor's state of mind with which Plato regularly endows him, ironically depreciates the validity of what he just finished saying, characterizing it as a "harangue" (*dêmêgoria*). He even imagines Protagoras himself as present and making the same criticism. The *scholium* has the following to say on the passage:

> The apparent refutation which Socrates has just now made of Protagoras when he mentioned the baboon he now calls a harangue. By means of irony he belittles his own argument so as to raise the young man's spirits.[22]

Similarly, Socrates' use of the tactful formulation, "You appear

17. Greene, 153, on *Gorgias* 489b.
18. See the *scholia* to *Alcibiades I* 120a (Greene, 97), *Gorgias* 471c (Greene, 154) and 491e (Greene, 154), and *Republic* 498e (Greene, 242).
19. Greene, 163, on *Gorgias* 500c.
20. Greene, 145, on *Gorgias* 473b.
21. Greene, 158, on *Gorgias* 495d.
22. Greene, 25.

(*kinduneueis*) not to be speaking the truth," was felt to reflect Socrates' wish to soften the impact of his refutation of one of Callicles' most cherished beliefs (489b).[23] Alcibiades is likewise corrected "unobtrusively" (*lelêthotôs*) by use of an example drawn from the world of the gymnasium. This would, as the *scholium* notes, have been familiar to him from childhood, the implied point being, it would seem, that such an example would have lead to an acceptance of Socrates' correction with a smaller resistance on Alcibiades' part (*Alcibiades I*).[24] Contrariwise, the intention to blame openly and without palliation is attributed to Socrates' at *Theaetetus* 181a, where he applies to the Heracliteans the epithet "in flux" or "flowing" (*rheontas*), characterized in the *scholium* as "a harsh and exceedingly satirical term."

Lastly, we may consider two instances in which the *scholia* argue that the words used by the interlocutors are a part of the subtle maneuverings of opponents engaged in dialectical contest. First, an example from the *Gorgias*. At 458a Gorgias suggests that the audience may perhaps be weary of having heard him speak at such great length, not only now with Socrates but earlier as well before Socrates came on the scene. Not so, says the *scholium*: Gorgias is afraid of continuing and "knavishly (*panourgôs*) puts the blame on those present for not wishing to continue."[25] So, too, at *Alcibiades I* 107e Socrates is understood to be shifting the responsibility for the formulation of the argument to his opponent Alcibiades, but unlike the "knavish" (!) Gorgias, he does so openly by means of a clever citation from Euripides' *Hippolytus* (352), where Phaedra, having led the nurse to utter the name of Hippolytus, similarly disclaims all responsibility for what the old servant has said.[26]

Further examples could be cited. What needs to be stressed, however, is that these observations, naive or obvious as they sometimes seem, are the expression of a principled determination on the part of the commentators to take Plato's text seriously, even at the seemingly meaningless level of conversational formulae or of turns of phrase which seem designed merely to keep the dialectic in motion. What the *scholia* at their best, in fact, embody is the important insight that Plato's dialogues are not merely representations of philosophical argument, but portrayals as well of complex and ambivalent human beings caught in dialectical confrontations which threaten their deepest conceptions of what they are and what they know, and which as a consequence provoke

23. Greene, 152.
24. See also Greene, 450.
25. Greene, 137.
26. Greene, 92.

a characteristic range of argumentative strategies and emotional responses by way of self-defense. The human struggle in Socratic dialectic has, to be sure, not gone unappreciated, but this appreciation is not so widespread that we should neglect to honor the commentators for their achievements.

I SHOULD LIKE to return to the question which was posed earlier about the possibility of reconstructing the interpretive system out of which these separate critical remarks arose. In studying these *scholia* it is often possible to formulate in some general way the critical assumptions which lie behind them, and I have tried to do this in several instances. These assumptions are in their way plausible enough — or at least have seemed so in many periods in Western history — but they are also extremely general. Many interpreters have agreed, to be sure, that in studying the text of a writer whose work seems to possess to some preeminent degree an organic or necessary character, to use Plato's formulation, it is methodologically proper to interpret the words and actions of such a fictive world in the light of what we take to be the author's larger structures of intention. However, such a formulation of the critical premises presumed to underlie the observations of the various *scholia* would still remain distressingly general in character. The unhappy truth in this matter — and here I suggest an answer to my earlier question — is that to anyone familiar with the Neoplatonic commentaries from which these observations were mined it is unmistakably clear that everything which is distinctly or explicitly Neoplatonic has disappeared virtually without a trace in the *scholia*. It is true that the critical methods of the *scholia* are *implicitly* Neoplatonic and that in this way they still bear the mark of their origin. Nonetheless it would be impossible, in my opinion at least, to recover from a study of the *scholia* any adequate or precise idea of the power, complexity and fruitfulness of the Neoplatonic theories or interpretation such as are found in the commentaries, especially those of Proclus, and in the *Prolegomena to Plato's Philosophy*. (This later work is not a commentary but, as the title suggests, an introduction to procedures for the interpretation of Plato's dialogues.)

The theoretical perspective which more than anything else endowed this system with its great power was the conviction that a dialogue of Plato was in a literal sense a microcosm analogous in all respects to the Great Cosmos in which we live. Everything that was true of the one was true of the other: when Plato wrote a dialogue, the small world he created bore within it the same complex structure of reality that the

Neoplatonists discerned in the universe itself. Therefore, to mention a single point, just as the world of phenomenal multiplicity was ultimately comprehensible for the Neoplatonists only as an emanation of Primal Oneness, so too, they argued, when we read a dialogue of Plato, no textual detail ever has meaning in itself. It must be viewed rather as a particular manifestation of the unity which pervades the whole dialogue, and this unity, at least in one influential formulation, they identified with Plato's conscious purpose or aim, his *skopos*. In the commentaries this is a key term, and it is usually an early order of business to determine what the *skopos* of the dialogue under discussion is. I have reserved this point for last because the only *scholium* which I have been able to find in which there is an explicit vestige of Neoplatonic exegesis has to do with precisely this interpretive principle. In the long introductory *scholium* of the *Sophist*, which perhaps derives from a lost commentary of Proclus, we are told[27] that in the view of the "great Iamblichus" the "purpose (*skopos*)" of this dialogue is to portray the "sublunary (i.e., terrestrial) demiurge." (This is in stark contrast to the heavenly Demiurge of the *Timaeus*.) The single, simple point I should like to make is that although the word "purpose" has, it would seem, a clear enough sense, it is also disquietingly evident that if all the Neoplatonic commentaries had disappeared — and with them the conceptual framework in which the word *skopos* was imbedded — we really would not have understood what the *scholium* was intended to convey, since the whole complex of exegetical concepts which gives this word its meaning would have perished along with the commentaries themselves.

■

27. Greene, 40.

V
Aristotle, Algazali and Avicenna on Necessity, Potentiality and Possibility

ARTHUR HYMAN

NECESSITY, POTENTIALITY AND POSSIBILITY formed one of the canonical problems of medieval philosophy, addressed by Muslims, Jews and Christians alike. * Since the discussion was independent of such privileged beliefs as the triune God for Christians, the created or uncreated nature of the Koran for Muslims and the eternity of the Torah for Jews, theologians and philosophers in the three traditions found it possible to share much common ground. At times there were direct influences, Jews borrowing from Muslims, and Christians from Muslims and Jews; but even where direct influences were lacking, philosophers in the three traditions developed similar positions. This is not surprising, since after all, they shared common questions and built on similar Scriptural beliefs and philosophic foundations.

While medieval philosophers discussed necessity and possibility, potentiality and actuality, and causality in purely philosophic works, such as commentaries on Aristotle and philosophic *summae*, they also discussed these notions in the context of the apparent conflict between Scriptural and philosophic teachings. Scripture taught that an omnipotent God created the world through the exercise of His free will, with purpose and out of nothing; while Aristotle had affirmed the eternity of the world, its contemporaneity with the prime mover or first cause, its lack of purpose other than being what, in fact, it was, and its functioning according to necessary laws. The Scriptural teachings gave rise to questions concerning the contingency of the world in its totality, a topic foreign to Aristotle's thought. More than that, miracles

* I am very pleased to participate in this tribute to Professor Paul O. Kristeller, a friend and colleague for many years, from whose seminal research all of us have learned so much.

recorded in Scripture placed in question the orderly working of nature and raised the possibility of the radical contingency of the natural order.

Confronted by these divergent views, dialectical theologians, called Mutakallimūn, and philosophers developed a variety of positions. While these differed in details and manifested subtle shadings, they may, I believe, be classified under three general headings. There were those who, critical of Aristotle, emphasized the power of God, maintaining that divine omnipotence requires that God's actions proceed from His undetermined will. God was not compelled, not even by His goodness, to create any world at all, and He could have created a world other than the one he did, in fact, create. While some proponents of the radical contingency of the world allowed for the orderly functioning of nature, there were others who extended radical contingency to that order as well. Just as the world in its totality was the product of God's unconstrained will, so were natural events and human volitional acts. The only answer to the question "what caused this?" is "the will of God." If experience disclosed some regularity in nature, it was the result of "habit" and if responsibility was to be assigned for human acts, it was because they were "acquired" by men. Arguing against necessary causes of any kind, proponents of this view admitted volition as the only cause. But even they did not maintain that God could do anything whatsoever; what is logically impossible was impossible even for God.

Then there were those who attempted to harmonize Scripture and philosophy. These granted the philosophers that the world is governed by necessary laws, though they went on the argue that God was the cause of this order. Miracles posed a problem, but at least some treated them as infrequent exceptions to the natural order. Against Aristotle, they argued for creation though, at times, this meant no more than that the world was contingent. Creation was not a temporal event, but an atemporal relation between God and the world. For these, the world in its totality was necessary in one respect, possible in another.

Finally, there were those for whom Aristotle was the final arbiter of truth. Scripture was primarily of political use, its laws designed to guarantee the orderly functioning of the state, its beliefs to assure the obedience of the masses. Penetrating the exoteric veil of Scripture, an accommodation to the imagination of the masses, they discovered an esoteric sense addressed to the intellect of an elite. For them, Aristotle's description of a world governed by necessary laws was true and Scripture had no role in their speculative quest.

In the present paper I shall undertake a three-fold task. Since the medieval discussion, both pro and con, rested on Aristotle's account of potentiality and actuality, causality and the eternity of the world, I shall begin with a discussion of these notions as they appear in the *Physics* and *Metaphysics.* Next, I shall consider the opinions of the Mutakallimūn, specifically those of the Ash'arites who, critical of Aristotle, defended the radical contingency of the world. For the exposition of their views I shall draw on Algazali's rather tellingly entitled *Incoherence of the Philosophers* and Maimonides' summary of their teachings in his *Guide of the Perplexed.* And finally, I shall consider one philosopher, Avicenna, who, in my opinion, is one of those who tried to harmonize Scriptural and philosophic teachings. (Since I am primarily interested in establishing a typology of positions, I shall forego discussing these thinkers in historical sequence.)[2] The opinions of the radical Aristotelians lie outside the confines of this paper.

"POSSIBILITY" OR "CONTINGENCY" in the sense of indeterminacy with regard to existence, so important for some medievals, plays a minor role in Aristotle's thought; and when he considers it at all, it is a potentiality of a certain kind. Hence, for Aristotle, "potentiality" (*dynamis*) rather than "possibility" is the basic notion. While he admits that potentiality and actuality extend beyond cases involving motion (*Metaphysics* 9.1),[3] it is a physical investigation, the analysis of motion or change, that yields these two related notions. Holding that without the reality of motion (*kinesis*) the natural order cannot be adequately explained, he defines it as "the fulfillment [actualization] of what exists potentially, in so far as it exists potentially (*Physics* 3.1, 201a, 10-11). Potentiality is then, first of all, a factor in motion or change. But more than this, it has a kind of reality of its own. "Being" *(to on)*, notes Aristotle, is divided in one way into substance and its accidents, in another it is distinguished "in respect of potency and complete reality, and of function" (*kata dynamin kai entelecheian kai kata to ergon.*) (*Metaphysics* 9.1,1045b, 32-34). What kind of being does potentiality possess? One intermediate between not-being and complete actuality.

While potentiality lacks the fullness of being that actuality has, in natural substances it is not indeterminate at all. Natural potentialities are

2. In temporal sequence they are: Avicenna (980–1037), Algazali (1058–1111), Moses Maimonides (1135 [1138]–1204).

3. For the Greek text I had reference to the edition of Immanuel Bekker, (Berlin, 1831; reprint Berlin, 1960); for the English translation to *The Works of Aristotle Translated into English,* eds. John A. Smith and William D. Ross (Oxford, 1908–1952).

directed toward definite effects, for to deny this is equivalent to admitting that everything can come from everything, and this is the position that Aristotle combats. In the ordinary course of events, fire burns, acorns become oak trees, and food is transformed into flesh and bones. If it happens at times that, for example, an acorn does not grow into an oak tree, it is an impeding cause, the absence of nourishment or sunlight, that keeps the acorn from becoming what it can be. But an impeding cause is a defect, and a defect does not point to a contrary potentiality, nor is it a sign that natural potentialities are undetermined.

While natural potentialities are determined toward one effect,[4] potentialities in men (and animals) may be determined toward contrary effects. (Aristotle used the term *dynamis* and its derivatives for both cases; but, to avoid linguistic confusion, I shall use the English term "possibility" for the latter case.) One of Aristotle's favorite examples for the second kind of potentiality, that is, possibility, is the proposition "the man [say Socrates] is seated." While it may be the case that Socrates is seated now, it is possible that he will not be seated at some future time. Observations such as these brought Aristotle to define what "possibility" means. Distinguishing between "possible" (*to dynaton*) and "impossible" (*to adynaton*), he defines the latter as that "of which the contrary is of necessity true," the former as that of which "it is not necessary that the contrary is false" (*Metaphysics* 5.12, 1019b, 23-30). It is impossible, for example, that the diagonal of a square be commensurate with a side since the proposition "the diagonal is *not* commensurate with a side" is necessarily true. But the proposition "the man [Socrates] is seated" is possible since its contrary is not necessarily false. It may be false, as a matter of fact, that Socrates is seated now, but "that Socrates is seated" may be true at some future time. These two definitions together with the definition of the "necessary" as that "which cannot be otherwise" (*Metaphysics* 5.5, 1015a, 35), became commonplaces later on, even for such anti-Aristotelians as Algazali.

If one then admits the existence of "possibility" in men (and animals) and at the same time denies its identity with "potentiality," to what cases does "possibility" apply? Aristotle answers this question when he maintains that "possibility" applies to acts that are the result of "desire" (*orexis*) or "choice" (*proairesis*), that is, voluntary acts

4. There are also instances in which the natural potentiality of an agent may be determined toward more than one effect. Fire, for example, depending on the conditions, may heat or burn.

(*Metaphysics* 9.2 and 5).[5] One may then go on to distinguish between potentialities of two kinds: nonrational (*alogoi*) and rational (*logoi*). Nonrational potentialities, such as the acorn's potentiality for becoming an oak tree, exist in all natural substances, living and nonliving alike and they are such that the agent produces a single effect. Rational potentialities, by contrast, such as Socrates' ability to sit or stand, are limited to men and are such that the same agent may produce contrary effects. It should be added that human voluntary acts, according to Aristotle, are not indeterminate, but require a principle of reason in the light of which a choice is made.[6] In medieval Muslim philosophy, as will be seen, the nature of choice became an important issue between the Ash'arites and the Aristotelians.

While all the four causes delineated by Aristotle (*Physics* 2.3) are required for an adequate explanation of motion or change, it is the efficient cause that is especially important for our investigation. Efficient causes, the agents of motion or change, exist in two modes: actual and potential. The builder prior to exercising his art is the potential cause of the house, but when he exercises his art he is its actual cause. Actual efficient causes, such as the builder engaged in building the house, are simultaneous with their effects, while potential efficient causes, such as the builder before he builds the house, are prior to their effects. However, once the effect has been produced (in our example, the house), it does not require a further cause in order to continue to exist. The cause of existence which some medievals required in order that anything may come to be as well as to continue to exist is absent from Aristotle's thought. Even in cases in which the efficient cause is always simultaneous with its effect — for example the nutritive soul with the acorn and the oak tree — it is only an efficient cause, not a cause of being.

From causality operative within the world, we turn briefly to causality and potentiality in the world in its totality. Arguing for the eternity of the world from the continuity of celestial motion (*Physics* 8, and *De caelo* 1.10-12), Aristotle affirms that there is no potentiality in the world considered as a whole. The world in its totality always was the way it is now and always will be. But even the celestial intelligences

5. What follows is based on these chapters. See also *Nicomachean Ethics* 3.1–5, esp. 2–3.

6. As will be seen further on (80–81), this was the view of the medieval Aristotelians. For a modern interpretation according to which acts of choice, while not necessary, have a cause, see Richard Sorabji, *Necessity, Cause and Blame* (Ithaca, NY, 1980), esp. chaps. 2 and 14–15. Sorabji also considers other interpretations.

and the prime mover differ from sublunar efficient causes in that they are always actual, never potential. "No eternal thing," states Aristotle, "exists potentially" (*Metaphysics* 9.8, 1050b, 6-28, esp. ll. 7-8). This Aristotelian principle became important to the medievals for showing how God can be the creator of the world without ever being a potential cause; for to ascribe potentiality to Him is to make Him subject to mutability.

WHILE ARISTOTLE CAME to causality, necessity, and potentiality and possibility through the analysis of an existing world and motions and changes within it, that is, through physical investigations, the Muslim Ash'arites approached these notions from theological, logical and epistemological perspectives. For them, the undetermined will of God is the only cause, necessity applies only to logical and mathematical relations, and potentiality and possibility are only epistemic, namely, the manner in which the human mind conceives the world and certain aspects of natural events.

Summarizing the opinions of these Muslim theologians, which he describes by the generic name "Mutakallimūn," Maimonides discovers their philosophic foundations in the teachings of Epicurus (*Guide of the Perplexed* 1.73).[7] With this philosopher they accept the existence of atoms, void, accidents inhering in atoms, and the atomic nature of time. But, whereas for Epicurus the atoms exist from eternity and the world comes to be by chance, for the Mutakallimun the atoms are constantly created by God and the world comes to be and functions through the free exercise of the divine will. It can readily be seen why this transformed Epicureanism is attractive to thinkers who affirmed the radical contingency of the world and who denied necessary laws of nature within it.

If, then, there is no regular order within the world, not even one imposed by God, everything that can be imagined is admissible for the intellect (*Guide* 1.73, tenth premise).[8] The element earth might move upward, men might have the size of mountains and elephants the size of fleas. This is the principle of "admissibility" (*tajwìz, ha'abarah*) which according to Maimonides, "is the main proposition of the science of the Kalam." If there is some apparent regularity in the natural order, it is

7. The following versions of the *Guide of the Perplexed* were used: Arabic, ed. Issachar Joel (Jerusalem, 1930–31); Hebrew, trans. Shemu'el ibn Tibbon, ed. Yehudah Even Shemu'el (Jerusalem, 1980–81); English, trans. Shlomo Pines (Chicago, 1963). For the reference to Epicurus, see: Arabic, 135; Hebrew, 170; English, 195.

8. What follows is based on this section. See also *Guide* 1.73, sixth premise.

the result of "habit" (*'ādah, minhag*). Invoking an analogy, Maimonides explains: it is the habit of the Sultan to pass through the marketplace on horseback, and he has never been seen to pass through the marketplace in any other way; but it is possible that the Sultan should pass through the marketplace on foot, and this is admissible.

While the principle of admissibility is meant to safeguard the omnipotence of God and His unconditioned will, there is one restriction: God cannot do what is logically impossible. God, for example, cannot will that a substance should exist without an accident, or that two contraries should exist in the same substratum simultaneously.

If then everything imaginable is equally possible, how can anything come to be? If, for example, it is equally possible that the sun be circular, triangular or square, what gives it its circular shape, and if a flower can be red or yellow, what makes it yellow rather than red? A "particularizing or differentiating principle" (*takhṣīṣ, hityaḥed, hityaḥadut*) is required, answer the Mutakallimūn. Since no independent necessary natural laws are operative within the world, only a volitional agent can bring one of the admissible possibilities into existence, and this agent is God. The Kalam system summarized by Maimonides had been developed extensively and with subtlety by his predecessor, the Ash'arite theologian Algazali, in his *Incoherence of the Philosophers.*[9]

Algazali begins his *Incoherence* with a defense of the creation of the world against the Aristotelian view that it is eternal. But even for Algazali creation is not a temporal process, that is, he denies that there existed infinite time at some point of which God brought the world into existence. Time, he holds, came into being with the creation of the world.[10]

If then creation does not mean that God created the world at some point in infinite time, how is creation to be understood? As Algazali sees it, to establish creation is to show the impossibility of infinite motion and time. To assume such infinity gives rise to puzzle cases concerning the infinite. Among the examples provided by Algazali

9. The following versions were used: Arabic, Algazel, *Tahafot al-Falasifat,* ed. Maurice Bouyges, S.J. (Beirut, 1927); English, in Arthur Hyman and James J. Walsh, eds. *Philosophy in the Middle Ages,* 2nd ed. (Indianapolis, 1984) (selections); and in Averroes' *Tahafut al-Tahafut* (The Incoherence of the Incoherence), trans. Simon van den Bergh, 2 vols. (London, 1954).

10. *Incoherence* 1.2 (Arabic, 52–53; English, van den Bergh, 38). See Michael E. Marmura, "The Logical Role of the Argument from Time in the *Tahāfut's* Second Proof for the World's Pre-eternity," *The Muslim World* 49 (1959): 306–14.

is one taken from planetary motion.[11] Observation discloses that the sun completes one orbit in one year, Jupiter one orbit in twelve years and Saturn one orbit in thirty years. It follows that for any distance covered by the sun in a given time (even infinite time), Jupiter covers one-twelfth of that distance and Saturn one-thirtieth of that distance. But if, as the Aristotelians hold, the world is eternal and the time of planetary motion infinite, it would follow that the three planets mentioned will traverse equal infinite distances in infinite time. Since then the two sets of distances covered cannot be both equal and unequal, and since observation has disclosed that they are unequal, it follows that the motions of these three planets must be finite and hence the universe.

Having argued for creation, Algazali must now show how the world can be created at all. For if, as he holds, everything is equally possible for God and if the world is said to be created by God's undetermined will, creation seems to be impossible. At best, the world came to be by chance. There is nothing in the possibility of the world that would require that one world rather than another should come to be, and there is nothing in the divine will that determines it to create this world rather than another. More than that, it appears that an undetermined will is unable to act at all. A "differentiating principle" seems to be required.

The search for a "differentiating principle" engages Algazali in a controversy with the Aristotelians concerning the nature of volitional acts and made him consider the problem which in the Latin tradition later on became known as that of "Buridan's Ass."[12] The Aristotelians had argued that a volitional agent confronted by two possible objects similar in all respects cannot make a choice.[13] Take, for example (and this is the example that Algazali cites), a thirsty man confronted by two cups of water similar in all respects. Unless there is a reason for taking one rather than the other, such as that one is more beautiful, lighter, closer to his right hand, he cannot take one of the two at all. If, then, it appears to be true that an undetermined volitional act is impossible and

11. *Incoherence* 1.1 (Arabic, 31–32; English, van den Bergh, pp. 8–9). See Lenn E. Goodman, "Ghazali's Argument from Creation," *International Journal of Middle East Studies* 2 (1971): 168–74; and William L. Craig, *The Kalam Cosmological Argument* (New York, 1979), 42–49.

12. Herbert A. Davidson, "Arguments from the Concept of Particularization in Arabic Philosophy," *Philosophy East and West* 18 (1968): 299–314, esp. 308–14; and van den Bergh 2, 1–2 and 17–18. See also Nicholas Rescher, "Choice Without Preference: A Study of the History and Logic of the Problem of 'Buridan's Ass'," *Kant Studien* 51 (1959–60): 142–75. For a recent discussion of this problem, see Edna Ullmann-Margalit and Sidney Morgenbesser, "Picking and Choosing," *Social Research* 44 (1977): 757–85.

13. *Incoherence* 1.1 (Arabic, 36–39; English, van den Bergh, 18–19).

there is some similarity between the human and divine will, even God requires some "determining principle" in order to act. From this argument it would seem to follow that the creation of the world through God's undetermined will would have to be abandoned.

Algazali counters the Aristotelians by saying that one could deny that there is a similarity between the human and divine wills by holding that while the human will requires a "differentiating principle," the divine will does not.[14] But this argument does not have to be invoked, since experience shows that there can be undetermined human volitional acts. In fact, the correct analysis of human volition shows that the will is its own "differentiating principle." "Will," states Algazali,[15] "is a quality which has the faculty of differentiating one thing from another."

To illustrate his view, Algazali offers an example of his own.[16] Take a man who has a strong desire for two dates, but cannot take both. Let it be assumed that such distinguishing qualities as beauty and nearness are absent. Experience discloses that even in the absence of a criterion for taking either, he will take one. It follows that no additional "differentiating principle" is required for a volitional act. If no such principle is required for man, none is required for God.

In countering an objection to creation, Algazali provides some understanding of his conception of possibility. If the world is created, an opponent can argue, then it must be possible before it came to be. But if this possibility is an ontological state, it must inhere in some underlying matter, since possibility cannot exist by itself. But if there is some preexistent matter, then the world cannot be created out of nothing, from which it follows that there is some constraint on the divine will.

Possibility, replies Algazali,[17] is not an ontological state but a judgment of the human intellect. Considering divine volition, one can only say that God wills whatever He wills without any antecedent possibility. It is only the human mind that considers the possibility (or impossibility) of things. Defining possibility and impossibility, Algazali writes:[18] "anything whose existence the intellect supposes,

14. *Incoherence* 1.1 (Arabic, 39–40; English, van den Bergh, 21).
15. *Incoherence* 1.1 (Arabic, 37, ll. 10–11; English, van den Bergh, 21). Maimonides defines: "…the true reality and quiddity of will means to will and not to will" (*Guide* 2.18; Arabic, 210; Hebrew, 262; English, 301).
16. *Incoherence* 1.1 (Arabic, 39–41; English, van den Bergh, 21).
17. *Incoherence* 1.4 (Arabic, 70; English, van den Bergh, 60) and what follows in the text.
18. *Incoherence* 1.4 (Arabic, 70, ll. 10–13; English, van den Bergh, 60).

provided no obstacle presents itself to this supposition, we call possible and, if there is such an obstacle, we call it impossible." Possibility (and impossibility), then, are not ontological but epistemic and they can only apply to man, not to God. For the case of God, one can only speak of acts of His undetermined will and their effects. Possibility is not predicable of God.

Just as Algazali assigns the creation of the world to the undetermined will of God, so does he assign events within it to the same undetermined will. Denying the necessary connection between cause and effect, he writes:

> The connection between what is customarily believed to be a cause and what is believed to be an effect is not necessary, according to our opinion, but each of the two is independent of the other. The affirmation of one does not imply the affirmation of the other, nor does the denial of the one imply the denial of the other; the existence of one does not necessitate the existence of the other, nor does the nonexistence of one necessitate the nonexistence of the other (*Incoherence* 17).[19]

Philosophers are led to necessary causation by faulty induction from experience. Having observed in a number of instances that fire burns cotton, they conclude that fire is the necessary acting cause of burning. But they are not entitled to this conclusion. All they know is that there is fire, cotton and burning. But nothing can be concluded about the causal agency of fire; God's will is the only cause.

Algazali argues his case not only from the theological principle of the undetermined divine will, but he brings an example from experience to support his view.[20] Suppose there is a man blind from birth whose eyes are covered by a membrane so that he cannot see, and who has never heard about the difference between day and night. If now the membrane is removed while it is day, his sight becomes sound

19. Arabic, 277; English, Hyman and Walsh, 283; van den Bergh, 316. For discussions of Algazali's account of causality, see I. Alon, "Al-Ghazālī on Causality," *Journal of the American Oriental Society* 100 (1980): 397–405; Lenn E. Goodman, "Did al-Ghazālī Deny Causality?" *Studia Islamica* 47 (1978): 83–120; K. Gyekye, "Al-Ghazālī on Causation," *Second Order: An African Journal* 2 (1973): 31–39; Michael E. Marmura, "Al-Ghazālī's Second Causal Theory in the 17th Discussion of his Tahāfut," in *Islamic Philosophy and Mysticism,* ed. Parviz Morewedge (Delmar, NY, 1981), 85–112; A. Marzouki, *Le concept de causalité chez Ghazālī* (Tunis, 1978); R. Shanab, Ghazālī and Aquinas on Causation," *Monist* 57 (1974): 140–50.

20. *Incoherence* 17 (Arabic, 280–81; English, Hyman and Walsh, 284–85; and van den Bergh, 317–18).

and a colored object which he now sees is present, he will undoubtedly think that the removal of the membrane is the cause of his perception of the colored object. Only when the sun sets will he understand that it is the presence of sunlight, not the removal of the membrane, that is the cause of sight. Similarly, philosophers think that the contact between fire and cotton is sufficient for burning, not aware that a volitional agent, the will of God, is the required cause.

If then not only the world in its totality but also events within it are the results of the undetermined will of God, would it not follow once again that everything can come from everything? In that case what would happen to the apparent regularity of natural events? In reply to questions such as these Algazali states, first of all (as we already saw in Maimonides' summary), that God cannot do what is logically impossible.[21] But more than that, while everything that is not impossible is possible for God, it does not follow that God must actualize all possibilities. Quite the contrary. Algazali states:

> God has created within us knowledge that He will not bring about everything that is possible and we do not assert that everything possible will necessarily come to be (*Incoherence* 17).[22]

In fact, God has arranged the natural order in such a way that things happen habitually time after time and He has constituted the human mind in such a manner that it can know this habitual order. Algazali writes:

> If something happens habitually time after time, its habitual course will be firmly rooted in our minds in accordance with the habitual past occurrence in such a way that it cannot be removed from our mind (*Incoherence* 17).[23]

It follows that habit is a feature of the natural order as well as of our knowledge of it. However, this opinion does not commit Algazali to some theory of abstraction according to which natural substances and events cause mental concepts and judgments. For just as God wills that

21. *Incoherence* 17 (Arabic, 292–94; English, Hyman and Walsh, 289–90; and van den Bergh, 328–29).
22. Arabic, 285; English, Hyman and Walsh, 286; van den Bergh, 324.
23. Arabic, 285–86; English, Hyman and Walsh, 286–87; and van den Bergh, 324.

nature follows a habitual order which is habitually known by the human mind, so does He will the parallelism between this natural order and our knowledge of it. Algazali, then, is not the radical occasionalist he is sometimes made out to be, but rather the critic of those who hold that the natural order is self-contained and functions according to inherent necessary laws.

IN THE FINAL SECTION of this paper I shall turn briefly to Avicenna as a representative of those who attempted to harmonize Scriptural and philosophic teachings. If one can trust his successors, this identification is not easy. Algazali considers him as one of the main spokesmen for the Aristotelianism he attacks, and he considers his views on creation, providence and resurrection heretical from the point of view of Islam.[24] Averroes, by contrast, accuses him of perverting Aristotle's views by intermingling them with Kalāmic teachings and undertakes to reestablish Aristotle's true teachings by cleansing Avicenna's opinions of their Kalāmic accretions.

Whereas Aristotle came to necessity, potentiality and possibility through the examination of the existing world, and Algazali through theological, logical and epistemological considerations, Avicenna follows a metaphysical approach. As a psychological starting point Avicenna invokes the example of the "flying man" (*Al-Shifā',* De anima 1.1).[25] Imagine a man created suddenly and perfectly, deprived of all sensory perceptions of the external world as well as of his body and its parts. A man of this description would still be aware of himself as thinking and he would know that he exists. Examining the contents of his mind, he would realize that "being" (*wujūd*) is the first notion known to the human mind. Reflecting further he would next

24. Algazali, *Al-Munqidh min al-Ḍalāl* (Deliverance from Error), ed. 'Abd al-Ḥalīm Maḥmud, 5th printing (Cairo, 1965), 100–107 (English, in Richard J. McCarthy, *Freedom and Fulfillment: An Annotated Translation of al-Ghazālī's al-Munqidh min al-Ḍalāl and Other Relevant Works of al-Ghazālī* (Boston, 1980), 76–77; and Hyman and Walsh, 272–73.

25. Arabic, *Avicenna's De anima,* ed. Fazlur Rahman (London, 1959), 16; French, in *Psychologie d'ibn Sīnā (Avicenna) d'après son oeuvre Aš-Šifa',* ed. and trans. Ján Bakoš (Prague, 1956), 12–13. See also, *al-Shifā's,* De anima 5.7 (Arabic, Rahman, 255; French, Bakoš, 181). The medieval Latin translation is found in Avicenna Latinus, *Liber de anima seu Sextus de naturalibus,* ed. Simone van Riet, 2 vols. (Louvain-Leiden, 1968–72). For discussions, see Soheil M. Afnan, *Avicenna, His Life and Works* (London, 1958), 150–52; A.-M. Goichon, *La distinction de l'essence et de l'existence d'après Ibn Sīnā (Avicenne)* (Paris, 1937), 13-15 and references there. The example of the "flying man" does not occur in the *Metaphysics of al-Shifā',* where Avicenna begins with the "intuition" of "being" (*Metaphysics* 1.5, beginning).

understand that certain distinctions are given concomitant with being. One such is that between essence and existence. He would be aware that while he is one substance, what he is, his essence, is ontologically distinct from the fact that he exists. More than that, he would understand that the mind can think essences without having the simultaneous judgment that they exist. In fact, the mind can think essences which do not exist now, even essences which can never exist. The question "what is it?" differs from the question "does it exist?" Existence, concludes Avicenna, is superadded to essence or, alternately, is an accident (*ʿaraḍ*) of essence.[26]

Another distinction concurrent with "being" and closely connected with that between essence and existence is that between "necessary existence" (*wājib al-wujūd*) and "possible existence" (*mumkin al-wujūd*). Since being is not a genus and, hence, necessary and possible existence not species, these two notions can only be described, not defined. Avicenna describes the possible as that which, considered in itself, does not have its existence (*wujūd*) by necessity nor is impossible, while the necessary is that which, considered in itself, has its existence by necessity (*Al-Shifāʾ,* Metaphysics 1.6).[27] Further reflection shows that that which is necessary through itself does not require a cause in order to exist, while that which is possible can only exist in the presence of a cause. To understand this contention better we must next examine Avicenna's conception of causality.

Avicenna begins his discussion of causality with the customary Aristotelian division of causes into four kinds. He writes: "The causes,

26. Interpreters have disagreed on how Avicenna's essence–existence distinction is to be understood and in what sense existence can be said to be an "accident" of essence. For literature on these topics, see Afnan, *Avicenna,* 115–21; Goichon, *La distinction,* 130–48; Parviz Morewedge, "Philosophical Analysis and Ibn Sīnā's 'Essence–Existence' Distinction," *Journal of the American Oriental Society* 92 (1972): 425–35; Fazlur Rahman, "Essence and Existence in Avicenna," *Medieval and Renaissance Studies* (Warburg Institute, London) 4 (1958): 1–16; Rahman, "Essence and Existence in Ibn Sīnā: The Myth and the Reality," *Hamdard Islamicus* 4.1 (1981): 3–14.

27. Arabic, ed. Georges C. Anawati et al. (Cairo, 1960), 37; English, Hyman and Walsh, 241; French, Avicenne, *La Métaphysique du Shifāʾ,* trans. Georges C. Anawati, 2 vols. (Paris, 1978–85), 1:13. For literature see Afnan, *Avicenna,* 121–26; Majid Fakhry, *A History of Islamic Philosophy,* 2nd ed. (New York, 1983), 152–56; Goichon, *La distinction,* 151–80; George Hourani, "Ibn Sina on Necessary and Possible Existence," *Philosophic Forum* 4 (1972): 74–86; G. Smith, "Avicenna and the Possibles," *New Scholasticism* 17 (1943): 340–57. Avicenna discusses the topics of this paper in several of his works, but I have limited myself to the discussions in *al-Shifāʾ.* For the medieval Latin translation of the *Metaphysics* of *al-Shifāʾ* see Avicenna Latinus, *Liber de philosophia prima sive scientia divina,* ed. Simone van Riet, 3 vols. (Louvain-Leiden, 1977–83).

as you have heard, are: form, element [matter], agent and purpose" (*Al-Shifā'*, Metaphysics 6.1).[28] Yet, for Avicenna, the efficient cause as cause of motion or change is not sufficient (as it was for Aristotle) to explain how something comes to exist. His metaphysical scheme in which essence and existence are ontologically distinct and in which everything other than God is possible in some fashion, requires, additionally, a "cause of being." Writes Avicenna:

> ...the metaphysicians do not intend by the agent the principle of movement only, as do the natural philosophers, but also the principle of existence and that which bestows [existence]... (ibid.).[29]

More than that, while, for Aristotle, something continues to exist once it has come into being, for Avicenna its possible nature requires that it have "a cause of being" in order to continue to exist. Avicenna also holds that a true cause must be simultaneous with its effect.[30]

Avicenna provides several examples illustrating his conception of causality, among them what occurs when a builder builds a house (*Al-Shifā'*, Metaphysics 6.2).[31] Ordinarily people think that the builder and his motions explain adequately how the house comes to be. But a more careful analysis discloses that this is not the case. The builder produces a series of successive motions — the movement of his arms and hands, the taking of the bricks, the placing of the bricks one upon another — but it is only with the cessation of the builder's motions that the house comes to be. While necessary for the production of the house, the builder is only an accidental or contributory cause.[32]

Continuing his analysis, Avicenna states that it is the coming together of the bricks that produces the shape of the house and this is the real essential cause of the existence of the house. But the bricks can only come together and remain together because they have a certain

28. Arabic, 257; English, Hyman and Walsh, 247; French, 2: 13. See Michael E. Marmura, "Avicenna on Causal Priority," *Islamic Philosophy and Mysticism,* 65–83; and Marmura, "Efficient Causality in Avicenna," *Islamic Theology and Philosophy: Studies in Honor of George F. Hourani,* ed. Michael E. Marmura (Albany, NY, 1984), 172–87.

29. Arabic, 257; English, Hyman and Walsh, 247–48; French, 2:13.

30. Arabic, 265, l. 1; English, Hyman and Walsh, 252; French, 2:19–20.

31. Arabic, 264–65; English, Hyman and Walsh, 252; French, 2:19–20.

32. *Al-Shifā'*, Metaphysics 6.2 (Arabic, 265; English, Hyman and Walsh, 252; French, 2:20). See also *Al-Shifā'*, Metaphysics 6.1 (Arabic, 263; English, Hyman and Walsh, 251–52; French, 2:18). For Avicenna on causality, see Fakhry, *History,* 149–53.

essence which is bestowed upon them by an incorporeal intelligence, the Agent Intellect also known as the "Giver of Forms" (*wāhib al-ṣuwar*). But even this intellect is only possible and requires a series of other intellects in order to exist. These intellects, in turn, are also only possible and ultimately require God, the being necessary through Himself, for their existence. From all this it follows that, while God acts through intermediary causes, He is required for anything — even the house — to exist. [33]

Avicenna's conception of necessity and possibility also influenced his interpretation of the Scriptural notion of creation. While there had been thinkers who interpreted creation as a temporal process, Avicenna rejects this account. There was no preexistent time at some point of which God created the world. Creation for Avicenna is an atemporal causal scheme of necessary and possible existence for which "emanation" (*fayḍ*) provides the model.[34] From God who is unique, immutable and necessary through Himself there emanates a first intelligence which is possible through itself, necessary through its cause. Reflecting upon itself as necessary, this intelligence emanates the soul of the outermost celestial sphere, reflecting upon itself as possible it gives rise to the body of this sphere, and reflecting upon its necessary cause (God) it brings forth the second intelligence. In similar fashion there proceed from the second intelligence the soul and body of the second sphere and the third intelligence. This emanative process continues until it comes to an end with the Agent Intellect and the sublunar world. It is only in the sublunar world that there exist transitory beings possible through themselves, that is, beings that may be or not be.

It has been seen then how Avicenna's cosmogony and cosmology are influenced by his metaphysical account of necessary and possible being. There is, first of all, God, the only being necessary through itself from which everything else proceeds and on whom everything depends for its existence. Next, there exists a series of celestial intellects, souls and

33. The passage cited at note 28 concludes: "...but also the principle of existence and that which bestows [existence], such as the creator of the world."

34. For Avicenna's account of creation as emanation, see *Al-Shifā'*, Metaphysics 9, esp. chaps. 6 and 7. See also Afnan, *Avicenna*, 126–35; Fakhry, *History*, 155–57; Goichon, *La distinction,* 201–84. It would appear then that while Avicenna is at one with Aristotle in holding that the world is eternal and exists simultaneously with God, he differs from Aristotle in holding that the world is ontologically contingent and depends for its existence on a cause necessary through itself. It appears further that he considers these ontological principles together with the account of emanation that he accepts as a justifiable interpretation of the Scriptural notion of creation.

bodies which, as eternal as God, are possible through themselves, necessary through another. While possible, they cannot not be. It is only in the sublunar world that there exist transitory beings possible through themselves, though even they, as we have seen, require a cause of being in order to exist.

■

VI
The Foliation of the Old English Life of Machutus

DAVID YERKES

HUMFREY WANLEY described London, Brit. Lib., MS. Cotton Otho A.viii* in his *Catalogus* of 1705 as a "Cod ex membr. in Quarto per diversorum manus, ac diversis temporibus scriptus, varios in se complectens Tractatus Latinos, qui ordine recensentur in Catalogo D. *Tho. Smithi,* ad quem lectorem mitto" (p. 232).[1] And Smith's *Catalogus librorum manuscriptorum bibliothecae Cottonianae* of 1696 has this entry for Otho A.viii (pp. 66–67):[2]

1. Vita S. Praxedis virginis.
2. Vita S. Mildrithae virginis: scripta tempore Scotlandi, Abbatis S. Augustini Cantuariensis, ut manu sua annotavit Joannes Joscelinus.
3. Textus translationis & institutionis Monasterii B. Mildrithae, cum miraculorum attestatione.
4. De obitu Bedae Cuthberti, ejus discipuli, epistola.
5. Vita S. Machuti, Episcopi Ventani, Saxonice.
6. Homiliae in natali S. Machuti, Christi Confessoris; ita titulus se habet, sed desiderantur ipsae, unico folio tantum relicto.
7. Versus de duodecim Apostolis.
8. Oratio in Hebdomada majori, feria quarta.
9. Oratio S. Gregorii Papae, cum interlineari glossa Saxonica.
10. Fragmentum orationis Bedae Presbyteri, itidem Saxonice interlineatum.

* This essay extends discussions from my edition *The Old English Life of Machutus* (Toronto, 1984) and articles in *Revue Bénédictine* 93 (1983): 128–31, *Notes and Queries* 31 (1984) 14–16, and *Manuscripta* 30 (1986): 108–11. *BHL* = *Bibliotheca hagiographica Latina antiquae et mediae aetatis* and *Supplementi*, ed. Société des Bollandistes, 3 vols. (Brussels, 1898–1911).

1. Humphrey Wanley, *Catalogus.*
2. Thomas Smith, *Catalogus librorum manuscriptorum bibliothecae Cottonianae* (London, 1696), 66–67.

11. Tabulae Chronologicae a Christo nato ad annum Ch. 1550. intersertis passim memorabilibus historicis.
12. Epitome privilegiorum Monasterii S. Augustini Cantuariae a prima fundatione per Regem Aethelbertum, ex variis Regum chartis. Vixit Auctor tempore R. Henrici v. Regis Angliae.

A Keeper's Room copy of Smith at the British Library includes the memo "Cod. *membr.* in 4to constans foliis 129. *lacer.* fol. 42." on a facing interleaf, as well as the following glosses in another hand in the catalogue itself after the manuscript's twelve items: "[1] F. 1. [2] F. 2. [3] F. 22. [4] F. 42. Joh: Joscelini manu. [5] F. 43. [6] F. 87. [7] F. 88. [8] F. 88.b. [9] ib. [10] F. 90.b. [11] F. 91. [12] F. 115." Thus, assuming no losses or misnumbering, the Old English Machutus — more precisely, the Old English translation of Bili's *Vita sancti Machutis* (BHL 5116) — comprised forty-four leaves, folios 43–86.

Losses and misnumbering came during and after the 1731 Cotton library fire, for since the British Museum had them inlaid and rebound in 1846 the entire MS Otho A.viii has comprised only thirty-four fragmentary leaves in tortured order and so scorched as to obscure any side as hair or flesh. Current folios 1-6 represent, with dwindling certainty, earlier fols. 3-4, 8, and 15–16 from the *Vita* and fol. 39 from the *Translatio sanctae Mildrithae* (*BHL* 5960–61), with fols. 2, 5, and 6 reversed. The other twenty-eight leaves, fols. 7–34, all come from the Old English Machutus.

Wanley had gone on to describe the manuscript's three Old English texts, glosses in Smith's items 9–10 and the

> *Vita S. Machuti Episcopi. Ventani* fabulosis narrationibus miraculorum repleta in Sectiones 80 distincta, & ante Conquaestum Angliae scripta. Longe amplior in hoc Codice habetur, quam in Latino opere de vitis Sanctorum gentium Britannicarum *Joannis Anglici* vicarii de *Tinmuth*; & in exscriptore ejus *Joanne Capgravio*.
>
> Incip. Vs gedafenað leofestan gebroþra...
>
> Expl. Ealle weorold gemæro. on his handa syndon Welan and wuldor and blisse. on him butan ende þurhwuniaþ. þæm is wyrþmynt 7 wuldor. on ealra weorolda weorold. AMEN.

Fol. 7r still has the heading (contractions expanded) "Incipit vita sancti machut[ ]uentani," the Old English incipit quoted by Wanley following on the next line, and collation with the Latin source — which means with the copy of Bili's *Vita* in Oxford, Bodleian Library, MS. Bodley 535 —

places all but the last of the extant leaves (asterisks denote reversed folios; Bodley 535 always has 31 lines a side):[3]

Otho A.viii	*No. of Lines a Side:*		*Bodley 535*
Folio No.	Now	Pre-fire	Folio/Line Nos.
7	22	24	65r/5-65v/18
[two(?) leaves missing]		24?	65v/18-
		24?	67r/1
9*	24	24	67r/1-67v/1
29	22	24	67v/10-68r/22
8*	21	24	68r/22-69r/2
32	24	24	69r/2-69v/10
24*	24	24	69v/10-70r/21
11	24	24	70r/21-70v/31
30	22	24	70v/31-71v/18
20	24	24	71v/18-72r/27
28*	24	24	72r/27-73r/13
25	24	24	73r/13-73v/30
27	24	24	73v/30-74v/10
18	22	22	74v/10-75r/20
17*	22	22	75r/20-76r/5
21*	22	22	76r/5-77r/9
23*	22	22	77r/10-77v/9
15	22	22	77v/10-78r/8
13	22	22	78r/9-79r/2
16*	22	22	79r/2-79v/3
33	22	22	79v/3-80r/28
[indeterminate gap]			
12*	23	24?	
19	24	24?	80v/11
31	24	24	80v/12-81r/18
22*	24	24	81r/19-82r/7
14	24	24	82r/7-82v/31
26	24	24	82v/31-83r/28
10	24	24	83r/29-84r/11
34	20	24?	

The explicit Wanley quoted — not in Bodley 535 — may have lain at the foot of fol. 34, either side.

Fols. 7 and 9 follow Bodley 535 closely, adding or omitting almost nothing. Each leaf translates about forty lines of Latin over a gap of about eighty. The next dozen leaves, through fol. 17, likewise add little and skip nothing longer than five lines at 73v/10–15 (Otho A.viii fol. 25v).

3. The second column counts the lines of text all or part of which survive. Fols. 29, 8, and 30 want their lines at the bottom of the page; fol. 12 wants its top line; fols. 7 and 34 may want top or bottom lines — which means 7v may have translated as far as line 19 or 20 of Bodley 535 fol. 65v. A corner of Otho A.viii fol. 9 got misbound as MS Otho B.x fol. 66, not reversed.

Thence some parting:

A. Fol. 21v fails to translate a passage of fourteen lines.
B. Fol. 23r–15r has a thirty-three-line Old English passage not found in Bodley 535.
C. Fol. 13v fails to translate a passage of fourteen lines.
D. Fol. 16v–16r has a thirteen-line Old English passage not found in Bodley 535.
E. Fol. 33v/4–8 corresponds in position but not content to 79v/27–80r/13.
F. Fol. 33v ends in the middle of a story, probably in the middle of a sentence, yet only the vestige of a chapter heading on the last line of the verso of fol. 19 corresponds to anything in Bodley 535: with 19v/24 [ ]*lis in uigilia* compare 80v/11 "DE TRIBVS EIVS MIRACVLIS IN VIGILIA." Fol. 19r continues without break the sermon begun on 12r, and the sermon begun on 12v ends at the top of 12r.[4]
G. Fol. 22r fails to translate a passage of seven lines.
H. Fol. 14v fails to translate a passage of eighteen lines.
I. Fol. 26v has a twelve-line Old English passage not found in Bodley 535.
J. Fol. 34 does not correspond to anything in Bodley 535 and leads neither in to nor out of any leaf.

Other witnesses of Bili's *Vita* (e.g., John of Tynemouth's abridgment that Wanley mentioned) supply additional Latin at fol. 23r–15r (B) but not at fols. 16 (D), 33 (E), 12–19 (F), or 26 (I), and not for fol. 34 (J). Fol. 34 may belong after 33 or 10, almost certainly not after 7. The scribe, who according to Ker wrote all twenty-eight extant leaves, used only the punctuation mark "․," for longer pauses on fols. 7 and 9–33, and only ";" on fols. 12–10.[5] Fol. 34 has the latter.

Eight consecutive leaves, fols. 18–33, have twenty-two lines a side. Taking the preceding eight, fols. 32–27, as a quire bequeathes an initial

4. Hitherto Neil R. Ker's *Catalogue of Manuscripts Containing Anglo-Saxon* (Oxford, 1957), Gwenaël Le Duc's *Vie de Saint-Malo* (Rennes, 1979), and my edition have all consigned fols. 12 and 19 to the limbo of fol. 34. Among the surviving leaves the spelling hwet occurs only at 12v/8 and 19r/4. Another correction to my edition: in the Old English glossary, on p. 127, line 3 from bottom, for *abstinendum* read *adstinentia*, and so in the Latin word index, on p. 167, line 13, delete *abstinendum*.
5. The lonely exceptions at 33v/5 and 19r/4 prove the rule. The former squeezes in where no pause belongs; the latter sits in an abnormally wide space, apparently over or before an erasure.

quire of six, fols. 7, [two], 9, 29, and 8 — unless that quire did not begin with the Old English Machutus. And in fact the Machutus scribe would have taken precisely two leaves to copy the text that once preceded the translation in Otho A.viii, Cuthbert's *Epistola de obitu Bedae* (Smith's item 4). I have put forward elsewhere other reasons for tying the texts: the spelling of Bede's Death Song in the *Epistola* seems to have agreed with the Machutus scribe's, the Machutus scribe may have gone on to copy the Latin *Homiliae in natali S. Machuti* (Smith's item 6) and both the *Epistola* and Smith's items 7–10 have a Worcester provenance. The Keeper's Room copy of Smith's catalogue assigns the *Epistola* one leaf, but torn, perhaps sign of missing a mate.

■

VII
On the Trail of a Lost Morningside Heights Manuscript*

ROBERT SOMERVILLE

> DR. ROCKWELL HAD AN EXTENSIVE RANGE of learning, and it was inevitable that he should interest himself in the Library.... He was an enthusiast for books. He startled and amused the Board of Directors by telling them that the growth of the Library necessitated much more room for the stack, and suggested that the entire dormitory (Hastings Hall) be emptied of students and turned into shelves for books.... His colleagues kept pressing him for current publications; but his major interest was in books already standard.[1]

Writing in 1954, the eminent retired president of Union Theological Seminary, Dr. Henry Sloane Coffin, thus characterized William Walker Rockwell. Dr. Rockwell had been professor of church history, and then from 1925 to 1942 the librarian, at Union, and was, in fact, still alive when President Coffin wrote those words, for he died in 1958 at age 83.[2]

Concealed beneath the remarks which follow is a much longer story of a modern search for a medieval Latin manuscript which once had been in Rockwell's personal library on Morningside Heights in New York

* This investigation has been dependent on information and suggestions supplied by many people — some no longer alive — over many years. It is, unfortunately, not possible to note here everyone who has answered a question or proffered advice, but in addition to those named in the notes below the following deserve special thanks: Myles M. Bourke, Paul A. Byrnes, Anne Clark, Dorothy Rockwell Clark, Ruth Rockwell Clark, Leon Festinger, Milton McC. Gatch, Robert T. Handy, Rudolf Hirsch, Paul Hoon, Paul Lehmann, Kenneth Lohf, Robert Maloy, Janet McKinney, the late Cyril C. Richardson, the late Katherine Lambert Richards Rockwell, Christiana Somerville, Richard Spoor, Beatrice Terrien, Sara and Samuel Terrien, and Kevin Trainor·

1. Henry Sloane Coffin, *A Half Century of Union Theological Seminary* (New York, 1954), 52–53.

2. For Rockwell and his career see *Who Was Who in America* 3 (Chicago, 1960), 737.

City. Whether or not that longer story could command general interest, its narration seems justified only if thus were defined a trail leading to the discovery of the lost item. That convergence is impossible, for the book remains hidden; but a brief account of the genesis of the quest and the results to date might be useful to historians of canon law, and publicizing the search could yield new leads.

The entry for New York City in the De Ricci-Wilson *Census of Medieval and Renaissance Manuscripts in the United States and Canada* is lengthy, and includes world famous libraries as well as small private collections.[3] When I came to Columbia University in the fall of 1969, those pages were consulted to gain some idea of the manuscript holdings in the City for the study of medieval canon law. The following listing for MS 5 in the library of Dr. W.W. Rockwell was of obvious interest:[4]

> Canon law, in Latin, books XIII-XXVI of an unidentified collection written after 1102 (cf.f.58r.); Pseudo-S.Bonaventura, *Summa de essentia Dei* (ff.120r.–123r.); varia (ff.123r.–124v.). Vel. (ca. 1200), 124 ff. (28x18 cm.). Imperfect at the beginning. Possibly written in the Rhine valley. xvth c. wooden boards and stamped calf, lined with 2 ff. of a xiiith c. antiphonarium.

The notice concludes with a five line heraldic description of unidentified arms — called fifteenth-century — on fols. 15r and 83v, and a final note indicating that the book was obtained in Geneva in 1922.

Thus began a seventeen-year search which still is in progress. Nothing is noted about the fate of any of Rockwell's manuscripts in the

3. Seymour J. De Ricci (with the assistance of W.J. Wilson), *Census of Medieval and Renaissance Manuscripts in the United States and Canada,* 3 vols. (New York, 1935–40). The entry for New York City is in the second volume.

4. De Ricci-Wilson 2:1812–14. Ten items are listed for Rockwell. MS 5 is the book at issue; the last three entries are miscellaneous, unspecified documents, fragments, and letters, in Greek, Italian, and Latin, and would be virtually untraceable. MS 1, an unbound fifteenth-century alchemical treatise, is lost; MS 2, a sixteenth-century *Processionale* from Liège, is Union Theological Seminary MS 79 (donated by Mrs. Rockwell c. 1970); MS 3, fifteenth-century prayers in Low German, is lost; MS 4, the Bedingfeld Book of Hours, remained in the Rockwell family until 1979, when it was sold at Christie's in New York (November 26, 1979; lot 493); MS 6, a signed letter from Philip of Hesse, is Union Seminary MS 78 (according to the files in the Seminary Library, "presumably" given before 1960 by Rockwell or Mrs. Rockwell); MS 7, a four folio *transactio inter Domnum Hyeronymum et S. Maurelium fratres de Jacobellis* (sic), Ferrara, Feb. 9, 1536, is lost.

Supplement to De Ricci-Wilson which was published in 1962, four years after Rockwell's death.[5] The cooperation of the Rockwell family makes it possible to conclude that the book no longer is in the family, and that no discernible familial records are at hand which might indicate what was done with it. With assistance from Union Theological Seminary it has been ascertained that MS 5 never was accessioned as part of the Seminary library collection. Portions of Rockwell's library were sold after his death to Midwestern Baptist Theological Seminary in Kansas City, Missouri, but no manuscripts were included in that consignment. Queries to other medievalists, scholars and former colleagues who knew Rockwell, and rare book and manuscript dealers and librarians in both North America and in Europe have provided many interesting asides, but not a trace of the codex.[6] Rockwell's daughter — Dorothy Rockwell Clark — is the only person alive who can be identified who ever saw the book, and her recollection is a childhood memory from Geneva in 1922, when her mother was upset with her father about the expenditure in acquiring it!

The notice in De Ricci-Wilson can be supplemented with three other partial descriptions of Rockwell MS 5. Two are very brief, and add little if anything to De

5. W.H. Bond and C.U. Faye, *Supplement to the Census of Medieval and Renaissance Manuscripts in the United States and Canada* (New York, 1962). Material has been assembled over many years at Houghton Library, Harvard, with the idea of revising De Ricci-Wilson, or at least preparing a new supplement. Dr. Laura Light of Harvard, and Dr. Jacqueline Tarrant of the Medieval Academy of America, kindly have probed in this material for hints about the fate of Rockwell's manuscripts, but have found nothing which would indicate what happened to those books.

6. Inquiries indicate that the book is not in the manuscript collections of the following institutions: the University of California, Berkeley; the University of Chicago; Columbia University; Duke University; the Getty Museum; Harvard University; the Huntington Library; the Library of Congress; University of California, Los Angeles; University of Michigan; the Morgan Library; the Newberry Library; the New York Public Library; University of Pennsylvania; Princeton University; University of Texas; Yale University. The author is grateful to librarians at these institutions for answering queries made at times in person, at times by mail. Responses to inquiries have been received from the following rare book and manuscript firms: William H. Allen (Philadelphia); Lucien Goldschmidt (New York City); Lathrop C. Harper (New York City); H.P. Kraus (New York City); Bernard Rosenthal (New York City, but as of the early 1970s San Francisco); Sotheby's (London); Laurence Witten (Southport, CT). Milton Epstein, who with his brother once owned the Ideal Bookstore on Amsterdam Avenue opposite the Columbia University campus, remembers Rockwell and his library, but is certain that Ideal never handled any of Rockwell's manuscripts. The indices for manuscripts covering the years 1933–1985 in *American Book Prices Current,* where such items sold at auction by major houses in this country should be listed, have been checked for all of the missing Rockwell manuscripts (see n. 4, above).

Ricci-Wilson. The third, in a lecture which Rockwell once prepared, is more elaborate and offers an idea about what this canon law book contained.

i. — On October 4, 1922, Rockwell wrote from Geneva to Dr. Henry Preserved Smith, then the Union Seminary librarian, as follows[7]:

> I am also going to negotiate for a manuscript of the 13th century, an excellent purchase for the Day Fund. You said I might buy things for that collection; and there is not much in that line in this Protestant citadel. This manuscript deals with canon law and liturgy. I think I can get it for a price well within the income of the fund.

The item in question clearly is the book which became Rockwell MS 5. Why Rockwell did not purchase it for the Seminary is a mystery. Perhaps it turned out to be too expensive for the Day Fund; perhaps Smith vetoed the suggestion; perhaps at the crucial moment Rockwell the bibliophile dominated Rockwell the Seminary representative and he bought the manuscript for his personal collection.

ii. — On January 24, 1924, the *New York Evening World* published a brief note under the lead, "Brings Old Latin Manuscripts to U.S.," with a sublead "Columbia Professor Returns With Valuable Mediaeval Writings."

> Two valuable medieval [sic] Latin manuscripts, one of the thirteenth century and one of the fifteenth, were brought from Europe to-day by Prof. W.W. Rockwell, Associate Professor of Church History at the Union Theological Seminary and member of the faculty of Columbia University, where he teaches paleography.[8] Prof. Rockwell, accompanied by his wife

7. This information was found in 1979 in the Union Seminary archives, Box ##21, by the author's then research assistant M. Beatrice de Bary, who surveyed holdings in the Seminary archives searching for information relevant to Rockwell's manuscripts.

8. The designation of Rockwell as professor at Columbia is not inaccurate, for he did teach courses in paleography there for many years: cf. R. Gordon Hoxie et al., *A History of the Faculty of Political Science, Columbia University* (New York, 1955), 85–86. A search in the *New York Times* for the same period revealed nothing comparable to the notice from the *Evening World,* although something was published in the *New York American.* In the lecture prepared for "Goddard Academy," to be considered, momentarily, Rockwell discusses the discovery of his alchemy manuscript (De Ricci-Wilson, Rockwell MS 1), which occurred during the same trip to Europe on which he purchased the canon law book. The lecture relates the following: "...after a voyage of ten days...I landed in New York. As there were very few people on board the purser of the *President Adams* turned reporters loose on me. I immediately tried to give them my views on foreign politics, but they weren't in the least interested....

> and their daughter, returned from a year's absence abroad....In an old bookstore in London he picked up the fifteenth century manuscript which contains an alchemist's recipe for making the "philosopher's stone," and also a recipe for converting baser metals into gold. Prof. Rockwell said the book store owner did not realize the value of the manuscript, and sold it "very reasonably". The thirteenth century manuscript deals with medieval church law. According to Sir Henry Martin, who lives at Paris, this is a most rare manuscript."

This clipping is clearly describing Rockwell MSS 1 and 5, as listed in De Ricci-Wilson, but the reference to "Sir Henry Martin" presents a problem. Henry Martin (1852–1927), was a renown librarian and manuscript scholar in France, who served as librarian of the important Bibliotheque de l'Arsenal, and who, in fact, was the editor of the great catalogue of over 8000 manuscripts in the Arsenal collection.[9] There is no indication that he was "Sir" Henry Martin; but in vain have old volumes of *Who's Who* and the *Dictionary of National Biography* been combed for a Sir Henry Martin who could be identified with the individual mentioned in the *Evening World* story. Probably the newspaper is in error and the expert cited — a reference which must have come from Rockwell — was, indeed, the noted manuscript scholar. Perhaps "Monsieur" was translated and then printed as "Sir."

iii. — A collection of Rockwell's papers is currently in the hands of his granddaughter, Ruth Rockwell Clark. In response to an inquiry she kindly searched through that material for information about her grandfather's manuscripts. She thus discovered a lecture which Rockwell composed for delivery at "Goddard Academy, Barre, Vt.," in June, 1932, titled "Hunting Manuscripts in Europe and America".[10] The text is composed of descriptions of some of his personal manuscripts plus reminiscences about how he came to own these books.

Suddenly...I thought of this manuscript about alchemy. I produced it; told them of its discovery. To my great surprise, I found myself on the front page of the *New York American* that evening, but with a headline not calculated to increase my reputation as a scholar: 'PROFESSOR BACK WITH OLD RECIPE FOR MAKING GOLD.'"

9. *Catalogue des manuscrits de la Bibliothèque de l'Arsenal,* 7 vols. (Paris, 1885–96).

10. The institution for which this address was prepared must be the forerunner of Goddard College in Plainfield, VT.

The discussion begins with the canon law manuscript, and very quickly the reader sees that the book in question was, at least in part, one of that group of medieval canonical manuscripts called penitential books.[11] It thus is possible to augment the meagre description from De Ricci-Wilson; and although Rockwell's comments are not systematic, and were prepared for a general audience and not for a group of historians, it is worth presenting here much of what he wrote concerning MS 5.

> In 1922 I spent six months in Geneva, Switzerland. One day while walking down La Grande Rue...I [entered] a second-hand book shop. The proprietor was a young man of perhaps thirty years, whose sister and brother-in-law I knew — a man of gentle spirit and of considerable learning.... I asked my usual...question, "Avez-vous des manuscrits?" He pulled out two or three in which I was not at all interested...and then he brought out the object which I now hold — Volume II of a work probably copied in the 13th century in a very neat, Gothic hand, with initial letters added, perhaps in the 15th century, by a scribe who incorporated into his design drawings of flamboyant Gothic architecture and certain intriguing coats of arms. I looked at it; I liked it; and I fell.
>
> I carried my prey back to my room and have spent a great many hours since, trying to find out just what I had bought...part of this manuscript is an old series of regulations as to the penance to be imposed on people for all sorts of sins. One section contains penitential canons...as follows:
>
> "Communion shall not be given at all to those who have committed homicide until they have performed penance;
>
> He who shall have arisen to smite a man, wishing to kill him shall perform three weeks of penance; if it was a clergyman who did this, six months of penance; if he merely wounded him, however, forty days. If a clergyman shall have attacked another clergyman he shall perform a whole year's penance (p. 100). A man who has killed his mother may not enter the church for an entire year, but when the year is up he may have permission to enter the church, but he may not receive communion until three years are up (p. 99v)."
>
> The 24th chapter of this work begins (p. 98) with the following table of contents....:

11. For penitential books in general see C. Vogel, *Les 'Libri Paenitentiales,' Typologie des sources du moyen âge occidentale* 27 (Turnhout, 1978).

"That no mere deacon is allowed to administer penances"

"The penance for those who burn down church buildings"

"Concerning those who talk in church during public worship"

"Concerning the different kinds of homicides"

"Concerning those who kidnap a man and sell him"

"Concerning the penance of those who wish to kill a man"

"Concerning those who take bad care of children"

"Concerning those who cut off babies' limbs"....

"Concerning those who dishonour or strike their parents"

"Concerning him who eats too much and those who rebel against the observance of Lent and of fasting"

"Concerning those who get themselves or others drunk and concerning gamblers"

"Concerning those who eat things which have been killed by wolves or dogs, and strangled things: What is to happen if you have eaten the food of Jews or of pagans, or have eaten or have drunk the blood of animals, or some liquor into which a mouse has fallen."

....This 24th chapter of the manuscript...lists in all nearly 90 kinds of sin, some of them described with such great frankness that the former owners of the manuscript have had a series of lines erased....

This Geneva Ms. is still unidentified. I have only Volume II, beginning with the last pages of Book 13 and continuing through the end of Book 26. There was never a title page, probably, to this second volume. The name of it, however, was indicated on the back of the binding, but the ink had so sunk into the ancient leather binding, put on about 500 years ago, that even the device of having it carefully photographed by the most skillful photographer in Geneva did not enable me to read what it said. I then resorted to chemicals. One day I took my manuscript out into the sun of one of the boulevards in Paris and applied the...reagent necessary to revive the ink. It is ammonium bisulphide.... Standing quite alone in the little strip of park in the center of this Parisian boulevard, I put on the drops of ammonium bisulphide, hoping to see the name of the manuscript appear; but alas! the ancient label of calfskin turned entirely black, so the secret of the name has probably perished forever. Having ruined the label, there was nothing to do but to try to identify it by its contents; I took the book to one of the most distinguished professors in the law school of the University of Paris, a member of the Academy of Inscriptions. With urbanity Professor Paul Fournier allowed

> me to see him at his apartment and he spent two hours examining with great interest this manuscript. He is the foremost authority in the world on the canon law from the ninth to the 12th centuries, and has seen over 200 manuscripts of the canon law between those dates. I was very happy when Professor Fournier told me that he never had seen a manuscript like mine; but he tempered my joy promptly by suggesting that it is probably a work which has had little circulation: — It may not be unique, but it was a real find, and a challenge to further investigation.

The extent to which this tantalizing account can help identify the collection in question remains to be investigated.[12] The lecture version of the manuscript's acquisition does offer, however, a very different perspective from the letter of October, 1922, to Union Seminary. No hint occurs here of Rockwell's suggestion that the manuscript would be a good purchase for the Union library; but obviously his purpose in the address was not congruent with a minute account of all details about how he discovered and then came to own the book. The Parisian expert consulted in this rendition was Paul Fournier, not "Sir" Henry Martin, but Rockwell probably made the rounds of a number of eminent scholars trying to identify his newly acquired treasure, and he no doubt consulted them both. Fournier's eminence in the history of canon law is undeniable, and, assuming that Rockwell's portrayal is accurate, his reaction upon studying the manuscript is interesting and adds to the mystery surrounding the book. But Fournier's writings after 1922-23 yield no mention of Rockwell's manuscript, neither in articles nor in the *Histoire des collections canoniques.* That silence is maintained in the famous article on "Pénitentiels" published in 1933 in the *Dictionnaire de théologie catholique* by Fournier's illustrious student Gabriel Le Bras, with whom he had collaborated in the *Histoire.*[13] It is,

12. The study of medieval penitentials is a thriving enterprise — cf. Raymund Kottje, "Die frühmittelalterlichen kontinentalen Büssbucher: Bericht uber ein Forschungsvorhaben an der Universität Augsburg," *Bulletin of Medieval Canon Law,* n.s. 7 (1977): 108–11. Despite the generalities and the English translations, circulation of the description given by Rockwell among scholars engaged in this research could yield an indentification for the collection.

13. Paul Fournier and Gabriel Le Bras, *Histoire des collections canoniques en Occident,* 2 vols (Paris, 1931–32); G. Le Bras, "Pénitentiels", *Dictionnaire de théologie catholique* 12.1 (1933), 1160–69. The vast bulk of Fournier's publications were prior to the 1920s. The article "Essai de restitution d'un manuscrit pénitentiel detruit," *Mélanges Mandonnet: Études d'histoire littéraire et doctrinale du moyen âge* 2, Bibliothèque

therefore, impossible to know what Paul Fournier truly thought of the codex.

There is an additional puzzle, now that something of the content of Rockwell MS 5 is known. Why is it not mentioned in John T. Mc Neill-Helena M. Gamber, *Medieval Handbooks of Penance*?[14] This work was published in the Columbia University Records of Civilization Series in 1938, and at the end of the preface the authors express their gratitude to Rockwell for placing the resources of the Union Library at their disposal and for suggesting books which might otherwise have been overlooked. Either they did not know about his canon law manuscript, or deemed it out of scope for their work. Until the book comes to light it is impossible to draw any sensible conclusions about the latter possibility; but perhaps Rockwell was hoping someday to identify the collection and publish his own study about it, and for this reason kept his penitential canons to himself.[15] Of course, soon after 1932 financial pressures might have forced him to sell the manuscript (but even so, McNeill-Gamer could have been informed about such a transaction and could have tracked down the codex in its new home).

THE PRECEDING PARAGRAPHS comprise everything which is known to the author about Rockwell MS 5. The search continues, and the present summary of the problem to date may help unearth new clues. It should

---

Thomiste 16 (Paris, 1930), 39–45, would have been an obvious place to mention a previously unknown penitential manuscript, but such does not occur therein.

14. John T. McNeill and Helena M. Gamer, *Medieval Handbooks of Penance,* Columbia University Records of Civilization, Sources and Studies 29 (New York, 1938).

15. Scholars who knew Rockwell have indicated that throughout his career he continued to read widely in church history, although he published very little scholarly work after the second decade of this century: cf. *A Bibliography of the Faculty of Political Science, Columbia University* (New York, 1931), 189–93, for an idea of what he wrote over the years. In this regard it is worth noting that not only do McNeill and Gamer list nothing about Rockwell MS 5, Lynn Thorndike and Pearl Kibre, *A Catalogue of Incipits of Medieval Scientific Writings,* rev. and augmented ed. (1937; Cambridge, MA 1963), and again Thorndike, *A History of Magic and Experimental Science,* 8 vols. (New York, 1923–58), make no mention of the alchemical treatise which was Rockwell MS 1. (For years Thorndike, as professor of history at Columbia, and Rockwell would have been neighbors on Morningside Heights in New York City, and their paths must have crossed on numerous occasions.) The incipits of this work, given in De Ricci-Wilson, do not occur in James Corbett, *Catalogue des manuscrits alchimiques latins*, 1–2: *Manuscrits des bibliothèques publiques de Paris* (vol. 1); *des départements français* (vol. 2) (Paris, 1939–51).

be pointed out, however, that according to the De Ricci-Wilson notice the manuscript concluded with a set of *varia*, prior to which is the pseudo-Bonaventura, *Summa de essentia Dei.* That text is yet to receive a critical edition, but attacking the question from this direction could prove fruitful.[16] Two further elements of the *Census* description deserve reiteration. The canon law collection is incomplete and neither titled nor ascribed — points which Rockwell's lecture confirm. The fifteenth-century coats of arms which are described in De Ricci-Wilson — and also noted by Rockwell — might be identifiable, and could thus offer entry points for trying to uncover the book in listings of manuscripts, although the Pseudo-Bonaventuran *Summa* is, at first glance, an easier referent of this sort. Furthermore, the description in De Ricci-Wilson noted that the codex' binding was lined with two folios of a thirteenth-century antiphonany, and this constitutes an additional factor which could attract scholarly attention to the book and prompt its emergence into the public domain.

"Lost" manuscripts do emerge. Optimism is essentially, and in that regard the engaging remarks in a personal letter received not long ago from Laurence Witten of Southport, Connecticut are appropriate. Mr. Witten is one of the country's most knowledgeable rare book dealers, and in response to my inquiry about Rockwell MS 5 he wrote in part the following:

> I feel sure I've never seen the W.W. Rockwell manuscript, but I hope I can acquire it and let you know one of these days. Things do have a way of going into hiding — often not very consciously — for a generation or two.

Mr. Witten's point can be pertinently illustrated in that at the time of this writing three separate portions of different medieval manuscripts have been discovered in the archives of Union Theological Seminary, including a portion of Vincent of Beauvais' *Speculum historiale* or a work closely related to the *Speculum*. The

16. Up to the present time this has not been the case. The author is very grateful to Fr. Conrad Harkins, O.F.M., the Director of the Franciscan Institute, St. Bonaventure University, and to Professor Servus Gieben, O.F.M.CAP, of the Istituto Storico dei Cappuccini in Rome, for helpful replies to inquiries. For the *Summa de essentia dei* cf. Balduinus Distelbrink, *Bonaventurae scripta* (Rome, 1975), p. 197, with a reference for the manuscripts to Fidelis a Fanno, O.F.M., *Ratio novae collectionis operum omnium...s. Bonaventurae* (Turin, 1874), p. 186. (Professor Uta-Renate Blumenthal, of the Catholic University of America, kindly sent to the author the relevant information from the latter work, which is a rare item in American libraries.)

immediate provenance of those pieces was Rockwell, who stored parts of his private library in the Seminary even after his retirement. None of these *membra disiecta* is represented in the De Ricci-Wilson list, perhaps indicating that Rockwell acquired them after the entry for his library in that work had been prepared.[17] It is not surprising that such things should emerge, especially at Union, for Rockwell's personal library was, by all accounts, enormous. At the end of his life he was in his ninth decade, and would not always have remembered clearly what he owned and where it was.

A portrait of W.W. Rockwell is housed today in the Library Director's office at Union Seminary. It is a large painting, and is signed in the lower lefthand corner "Aliceque Cann Boscawitz, 1939." The subject, a rather youthful figure for a man of 65 years, is seated and dressed in doctoral robes. His face appears cordial and intelligent but not without a certain reserve. In his hands he is holding a book bound in a blue binding, and a viewer readily can see that it is an illuminated manuscript, obviously the Bedingfeld book of hours (Rockwell MS 4, in De Ricci-Wilson). Perhaps the canon law manuscript was on a table or a bookcase in the background.[18]

■

17. None of these fragments could be nos. 8–10 in De Ricci-Wilson (cf. n. 4 above).

18. Postscript. On the basis of the description of Rockwell's manuscript given in this article, Professor Stephan Kuttner has identified the work as the *Candela* of Gerlard of Besançon. Cf. Stephan Kuttner, "Gerlard of Besançon and the Manuscripts of his *Candela*: A Bibliographical Note," *Paradosis: Studies in Memory of Edwin A. Quain* (New York, 1976), 71–84.

VIII
Decameron I/7: The Literary Space of a Text

KARL-LUDWIG SELIG

"es ist ein weites Feld"
Fontane, *Effi Briest*

BOCCACCIO'S *DECAMERON* is a crisis text, a pivotal and keystone text, a watershed text in and for the history of literature, for the history of storytelling and the narrative. It is the codification of a great literary tradition and genre: the novella. A new impulse is given to a textual tradition. As a text, and as it represents a textual system, it is of course an artifice, a simulacrum, a literary, intellectual, artistic and verbal game — with a remarkable ludic sense and spirit — a *Wortkunstwerk*. It is a game played by an author/artificer/creator/critic, who often permits the text to give off signals, signs, which give or set off clues or intimations pertaining to the strategies and processes of his textual system. In this respect, as the text calls attention to itself and its processes and logistics, it is a self-conscious text, a text which discusses itself, where the text itself is under scrutiny.

Incorporating all previous traditions and conventions pertaining to relatively simple narrative structures and forms — texts of an Oriental provenance and some transmitted from Classical antiquity, texts of an oral and/or written tradition, and structures such as the *exemplum*, apologue, fabliau and such highly reductive forms as the proverb — the *Decameron* presses the world of the narrative to new possibilities, to new dimensions and frontiers, to new spheres and spaces of the literary and artistic imagination. The text in its totality and organicity is in its own way an epic; it is an encyclopedic text, which encapsulates an immense world. In its epic scope and in contrast to some other (contemporary) texts and literary and artistic traditions and endeavors or articulations, of which the text informs a decided and remarkable but not necessarily a polemic awareness or consciousness, there is a decided stress on the mercantile, or worldliness, emphasized and intensified by a consciousness, an awareness of a particular centrality — the pivotality of history, of contemporary life, of a contemporary world. And while the text ranges over many terrains, spaces — ultimately literary spaces — and many epochs and periods of time, one must not forget the importance, the centrality of Florence, so often referred to as "our city." As the text is set against a disorder outside — but in contemporary history and initially identifiable geography which is ultimately transformed into literary and artistic geography and space — it establishes an order by way of the text and the textual enterprise — by way of a work of art. Procedure and code are of fundamental importance and, furthermore, it must be indicated that the texts and the storytellers are in relationship to each other; they are in competition with each other. What dominates — what prevails — is the art, the magnificent craft of the text — a verbal tour de force. The word and verbal wit and the strategies of verbal wit are fundamental; they are the tools of the craft and the artifice of the text; the *Wortkunstwerk* is a tour de force; there is stress on virtuosity.

LET US NOW SEE how these, what could be called general observations and premises or postulates, are illustrated or exemplified in one very concise and reductive structure, in one particular novella. It is the story of Bergamino and the Can Grande della Scala. Let me briefly summarize the "story": Bergamino arrives at the court of the Can Grande della Scala; for some reason he is not received with much hospitality; in order to teach the Can Grande a lesson, Bergamino tells a story about Primas and the Abbot of Cluny.

In my analysis of this novella, as it exemplifies the craft and virtuosity of the text — and is quite symptomatic of many central aspects and characteristics of the total text — but taking of course into account the givens of the text and what I like to call the literary and artistic problems generated by the text, it can be shown that, while observing certain fundamental traits and facets of the novella (and therefore the critical challenges of the genre, for this genre represents a tradition which on the one hand is observed and respected but which is also challenged and which is also open to innovation), such as limited "story" element and even relatively limited narrative or fictional time, the "story" revolves essentially around one central episode/incident, limited number of personages or cast of characters — nevertheless an "extraordinariness" is or can be attained, as the text reaches out to new and wider spaces, terrains, literary geography and traditions and encapsulates these elements into the literary space of the text. It pertains to the matter of the extension of the textual parameter by the manner of the manipulation of the artistic articulation of explicit and implied space and time and possibly explicit and implied text(s). I must also take the liberty of stating here, as an additional postulate, that I consider the text a textual system of information; in a novella, given the compactness of the form and structure, this system has for artistic reasons significant but challenging limitations, and the release of information, of facts — and therefore similarly or correspondingly the withholding of information and facts — has to be handled in a particularly eclectic way and is a key factor or element pertaining to the strategies, strategems, and logistics of the text.

THE TEXTUAL SITUATION of the seventh novella of the first day, in contrast to many novelle in the *Decameron* which are centered on Tuscany or set in Florence, takes place in Verona at the court of the Can Grande, who is holding and celebrating festivities. It revolves around and deals with a well known person and therefore someone with a very particular resonance, who had a wide reputation and an important position in history, who is significantly alluded to in other (contemporary) texts, such as the *Divine Comedy*; and it is furthermore well known that he was considered, in fact hailed, as a kind of messianic personage, a figure of hope, a possible unifier in and for a divided country, a country torn apart by wars and civic strife. He had also an anecdotal fame as a good and generous host, and one must recall that Boccaccio includes many "legends" or anecdotes about famous men, e.g., King Charles, Giotto, in his text and which served as the basis or kernel for the/a novella.

As our text, the *Decameron*, is the codifying text of the novella tradition, nevertheless I must insist once more in order to focus on the thrust of the individual novella which I am examining, that there are in the *Decameron* many textual instances which indicate an awareness and consciousness of other textual and generic traditions and forms of art; it is a brilliant example of the function of an (implied) intertext. Not only is Can Grande praised in I/7, but he is compared and paragonized to Frederick II, a model monarch and ruler of a model court which also represented a great intellectual center of a somewhat earlier period and at the other end of Italy; the extended spatial field of the text is effected by an allusion and reference to a north-south axis (Verona-Palermo).

CAN GRANDE, given his reputation as a splendid host, arranges festivities, which in their own way represent various forms of art and artistic activity. Bergamino, a fictive personage but known according to the text as a wit (*parlatore*) and popular entertainer at various courts, and aware of the reputation of the Can Grande, is somehow upon arriving at the latter's court not received — for some quirky reason — with the usual hospitality; he stays at an inn and is obliged to pay for his expenses with three robes which he had received from another patron (*tre belle e ricche robe*). What is brought into opposition and confrontation is the well known historical personage and a popular but fictive character; this admixture is important (proto-) paradigmatically for historical fiction and/or for texts anchored in history; furthermore the mention of the festivities relates the text to important cultural traditions.

One day the Can Grande, finding Bergamino in a rather sad and gloomy mood, teases him and is curious why he is so melancholy. Bergamino retorts in a brilliant way; he tells a story which is relevant to his own state:

> ...allora, senza punto pensare, quasi molto
> tempo pensato avesse, subitamente in acconcio
> de' fatti suoi disse questa novella.

It is a test of Bergamino's wit, but even more so the brilliance of the craft of the text is indicated, as the artificer, the creator of the text, manipulates Bergamino to feign spontaneity; he has the proper cool; of course he had been waiting for this moment; he had been prepared all day. The text is an artifice; it is a simulacrum, and as there are synergetic layers which function in and for the text, and as the text runs on a number of inter-operative tracks, one essential and pivotal one is suggestively the track of the theory of the text. Spontaneity, the simulation of spontaneity, improvisation, the simulation of improvisation all hint at important and desirable ploys and aspects of the craft of the text.

Bergamino tells a reflector story, a parallel story fitting his own state of affairs. In this way, and also bearing in mind that the "stories" contain elements of the *exemplum* and the anecdote, the Can Grande/ Bergamino story forms, so to speak, the frame story, and at the same time with the reflector story, we have ultimately a unified double story. Bergamino tells the story about Primas and the Abbot of Cluny (the latter's reputed wealth is also mentioned in *Decameron* X/2). By means of the reflector/parallel story the space and the temporal sphere is once more extended; to put it another way, the actual locus and tempus of the core narrative, which also functions as a frame story, is extended as the text reaches out to another cultural and intellectual center, to another period, and to another literary tradition. The text now informs not only the axis Verona-Palermo but also Verona-Paris-Cluny, and not only the polar Can Grande-Bergamino, but also the binary Abbot of Cluny-Primas.

Primas, who is often associated with the tradition of the wandering poet, with the goliards and the goliardic texts, exists in a somewhat legendary and mythified tradition. But he is known for verbal power, eloquence; he stands for grammar; at the core is language; the use and function of language are central, pivotal, critical; language is at the core of everything, and in this instance, at the service of wit — and to survive. Primas has heard about the Abbot's reputation and wanted to partake of his hospitality, his table, his generosity. Somehow the Abbot finds a stranger at his table whom he does not recognize and who looks scruffy; the term *ribaldo* is mentioned twice. Parallel to the three robes of Bergamino, Primas provided himself with three loaves of bread, which he is forced to consume while the Abbot leaves him without any food. Ultimately, the latter recognizes his fault, his moral blemish, and here the thrust of the exemplary element and the framing device/aspect of the text permit a fusion of the literary tradition/convention of the *exemplum* into a novella. And the anecdotal matter/element is elaborated upon by the "story" or a more extended narrative, for in this way the "story" or narrative has what can be considered an exegetical function, as it permits the permutation or shift from a very simple structure to a somewhat or slightly more complex form and structure.

The Decameronian novella can be brilliantly concise and sparse, and Can Grande does not need to hear more; he repents his miserliness; the great man has learned a lesson, and he regales Bergamino:

> gli fece le sue tre robe restituire, e lui nobilissimamente d'una sua roba vestito, datigli denari e un pallafreno, nel suo piacere per quella volta rimise l'andare e lo stare

just as the Abbot had parallelly given Primas

> denari e un pallafreno, nel suo arbitrio rimise l'andare e lo stare.

No more has to be said; Can Grande can take a hint; he has understood *quel bastone che tu medesimo hai divisato;* the implied, the unsaid will do; it pertains to the artistic function of silence. Can Grande's reputed and tested exemplariness is restored to his true exemplariness, the qualities and attributes of a great man, of a potentially great leader — hopes and ideals which are intimated by the magic suggestiveness of a great work of art.

■

IX

Appropriate Enough: Telling 'Classical' Allusions in Chaucer's *Canterbury Tales*

ROBERT W. HANNING

THROUGHOUT HIS WORKS, Geoffrey Chaucer appropriates for his own, varied purposes elements from the writings and traditions of classical antiquity. Many scholars have assessed Chaucer's "debt" and "borrowings," both general and particular, and a recent bibliography of Chaucer "source studies" testifies, by its girth and its many relevant categories, to the thoroughness of such labors.[1] Nonetheless, it is still possible to shed new, and perhaps useful, light on Chaucer's poetic relationship to pre-Christian culture by examining a few instances in his poetry — primarily the *Canterbury Tales* — where allusions to classical texts, or to medieval recreations of pagan life and times, form part of his overall narrative strategy. Slight as these references may appear, they contribute to the ironic texture and comic effect of Chaucerian poetry, and as such deserve our scrutiny. In this case, observing how Chaucer builds on his cultural heritage also serves as a fitting gesture of salutation and praise for a colleague, Paul Oskar Kristeller, whose activities in reclaiming the past, and in guiding others engaged in the same laudable enterprise, have provided inspiration to so many.

Chaucer was never shy about appropriating episodes and set pieces from classical poets, as his raids on Ovid and, to a lesser extent, Virgil

1. See Lynn King Morris, *Chaucer Source and Analogue Criticism: A Cross-Referenced Guide* (New York, 1985). Among recent books dealing with Chaucer's use of the classics, John Fyler's *Chaucer and Ovid* (New Haven and London, 1979) stands out for its critical sophistication. Alistair Minnis, *Chaucer and Pagan Antiquity* (Cambridge and Totowa, NJ, 1982), has the specific aim of explicating, and placing in its late medieval context, Chaucer's attitude toward the beliefs, virtues, and failings of those pagans who lived before Christ's incarnation. For an introductory (and somewhat flawed) treatment of the subject, see Bruce Harbert, "Chaucer and the Latin Classics," in D.S. Brewer, ed., *Geoffrey Chaucer,* Writers and Their Backgrounds Series (London, 1974), 137–53.

in what are probably his earliest poems — *The Book of the Duchess* and *The House of Fame* — demonstrate.[2] But he also enjoyed subtler allusions to this material, by which he enriched the resonances — frequently ironic — of particular passages. I have recently indicated one such allusion in the fourth book of his great love-tragedy, *Troilus and Criseyde*, where the protagonists meet, in Criseyde's chambers, for the first time after receiving the dreadful tidings that she is to be sent to the Greeks in exchange for the captive Trojan noble, Antenor.[3] As the lovers embrace, they weep profusely, and the narrator comments that

> Tho woful teris that they leten falle
> As bittre weren out of teris kynde,
> For peyne, as is ligne aloes or galle —
> So bittre teris weep nought, as I fynde,
> The woful Myrra thorugh the bark and rynde —
> That in this world ther nys so hard an herte
> That nolde han rewed on hire peynes smerte.
> (4.1135–41)[4]

The point of establishing a comparison with Myrrha — whose story Chaucer would have known from the tenth book of Ovid's *Metamorphoses* (lines 298–502) — is that Myrrha achieved infamy within the world of classical legend and myth by her inability to stay away from her father, and this characteristic makes her an ironic "type" of Criseyde who manifests the same inability, albeit for reasons very different from that of her incestuous antecedent.

A similar, pointedly ironic — and, I believe, hitherto unnoticed — allusion to a classical myth occurs in the "Merchant's Tale" of *The Canterbury Tales*. As January, the foolish, sensual old Lombard knight, ponders marriage, he explains in considerable detail to his retinue

2. See, for example, Fyler, 23–81; and R. W. Hanning, "Chaucer's First Ovid: Metamorphosis and Poetic Tradition in *The Book of the Duchess* and *The House of Fame*," in Leigh Arrathoon, ed., *Chaucer and the Craft of Fiction* (Rochester, MI, 1986), 121–63. The subject of Virgil's presence in *The House of Fame* accounts for 46 entries in *Chaucer Source and Analogue Criticism* (see the preceding note).
3. See "'I shal finde it in a maner glose': Versions of Textual Harassment in Medieval Literature," in Laurie A. Finke and Martin B. Shichtman, eds., *Medieval Texts and Contemporary Readers* (Ithaca, NY, and London, 1987), forthcoming. In *Chaucer and the Poets* (Ithaca, NY, and London, 1984), 98–101, Winthrop Wetherbee discusses this passage but ignores the point I am making here, concentrating instead on the parallels between Myrrha's situation in Ovid and that of Troilus in Chaucer's poem.
4. All quotations from, references to, and abbreviations of the parts of Chaucer's works follow *The Complete Poetry and Prose of Geoffrey Chaucer,* ed. John H. Fisher (New York, 1977).

why he will never take as a wife an old, much married woman (such as the Wife of Bath, with whom the Merchant is obsessed throughout his tale).[5] Such would be January's revulsion from one of "thise olde wydwes" (MerchT 1423) that he could take no sexual "pleasaunce" with her; as a result, he avers, "... children sholde I none upon hire geten,/ Yet were me levere houndes had me eten/ Than that myn heritage sholde falle in straunge hand" (1437–40).

January's exclamation constitutes a transparent reference to the story of Actaeon, the Theban hunter who (by either design or accident, depending on the version), sees the goddess Diana naked at her bath and is punished for his trespass by being attacked and eaten (or at least torn apart), by his dogs.[6] As the Merchant's Tale moves toward its bitter denouement, January's words, with their suggested equation of himself and Actaeon, become doubly ironic. First, Damyan, January's squire, falls in love with his master's wife, May, and is rendered bedridden by his passion until May agrees to take pity on him, whereupon he arises and "to Januarie he gooth as lowe/ As evere dide a dogge for the bowe" (2013–14). That is, Damyan's submissive behavior makes him seem like one of January's hunting dogs, even as he turns on his master by planning to cuckold him.[7] And when the cuckolding takes place, in January's own private love garden, the old knight, long blind but now given sight, out of sympathy and solidarity, by the god Pluto, is suddenly, like Actaeon, put in the position where he sees something he "should not," i.e., his wife and his squire making love in a tree. (May then convinces her doting husband that he has seen what he

5. Compare, for example, MerchT 1427–28: "For sondry scoles maken sotile clerkis:/ Womman of manye scoles half a clerk is," and WBProl 44c–f (lines not in the earliest extant manuscripts of *The Canterbury Tales*; apparently a late addition to the text intended by Chaucer to establish a tighter fit between the two tales and tellers): "Diverse scoles maken parfyt clerkes,/ And diverse practyk in many sondry werkes/ Maketh the werkman parfit sekirly;/ Of fyve husbondes scoleiyng am I." See also the ironic appeal to the Wife as an authority on marriage by Justinus (MerchT 1685–87), and the Alisoun-like diction and opinions of the goddess Proserpina in January's garden (2264–2310).

6. See Ovid, *Metamorphoses* 3.138–252; Ovid insists that Actaeon was a guiltless victim of Fortune, which brought him unwittingly to the goddess's ablutions. By contrast, several later retellings of the story provided Actaeon's lust as a motive for his (deliberate) violation of Diana's place of refuge. Erwin Panofsky, *Problems in Titian, Mostly Iconographic* (New York, 1969), 154, n. 42, mentions Nonnus (in his *Dionysiaca*) and Hyginus (in his *Fabulae*) among the proponents of this latter version.

7. Cf. the irony of January's avowal to May that "A man may do no synne with his wyf,/ Ne hurte hymselven with his owene knyf" (MerchT 1839–40), in light of the fact that Damyan, who will indeed "hurt" January, "carf biforn the knyght ful many a day" (1773) as part of his duties as a squire.

should not in the very different sense that his eyes have deceived him in their first moments of regained sight [2380–83, 2387–89, 2396–2410].) [8]

The "Miller's Tale" offers two particularly delightful instances of Chaucer's creative use of "classical" allusion, both of which contribute to the richness and complexity of this most celebrated of medieval English fabliaux. In fact, one of the references is to a story of medieval origin but dealing with the pagan past, and therefore possessing for Chaucer and his contemporaries the same kind of authority — and limitations — as materials originating in antiquity.

The first allusion comes in the Prologue to the Miller's tale, at a moment of great import both for the fictional storytelling contest and for Chaucer's poetic project.[9] The Knight has finished his story and received general approbation (especially from those pilgrims of higher social rank); the Host pronounces himself pleased with this first entry in the story-telling contest over which he is presiding, and asks the Monk (another "gentil" pilgrim) to speak next. Here the Miller breaks in, to challenge both the Host's authority and his agenda, and as he does so, the narrator describes him as "the Millere, that fordronken was al pale,/ So that unnethe upon his hors he sat" (MilProl 3020–21). These two lines constitute a close paraphrase of Ovid's description, in the first book of his *Ars amatoria,* of Silenus, the companion and tutor of Bacchus: "Ebrius, ecce, senex pando Silenus asello/ Vix sedet, et pressas continet ante iubas."[10] The allusion suggests that Chaucer has

8. The story of Actaeon is not the only myth evoked ironically in MerchT. The same passage that contains January's reference to being eaten by his hounds also has him desire a young wife because "a yong thyng may men gye/ Right as men may warm wex with handes plye" (1429–30). This allusion to the story of Pygmalion, who creates, and subsequently falls in love with, a statue of a beautiful woman, finds its ironic fulfillment in the warm wax used by May, January's "creature," to take an impression of the key to January's love garden, as part of her and Damyan's scheme to consummate their love (2116–21). (Hope P. Weissman analysed the relevance of the Pygmalion myth to the "Merchant's Tale" in an unpublished paper some years ago.)

9. I have refered briefly to this allusion in an earlier essay, "The Theme of Art and Life in Chaucer's Poetry," in George D. Economou, ed., *Geoffrey Chaucer: A Collection of Original Articles* (New York, 1975), 29.

10. Ovid, *Ars amatoria* 1.543–44; I cite from the Loeb Classical Library edition, *Ovid: The Art of Love and Other Poems,* trans. J. H. Mozley (1929; repr., London and Cambridge, MA, 1957). Mozley translates these lines, "lo! drunken old Silenus scarce sits his crookbacked ass, and leaning clings to the mane before him." That Chaucer was thinking of this depiction of Silenus is, I believe, corroborated by the passage in the Prologue to the "Manciple's Tale," mentioned in the next paragraph, where the Cook becomes so drunk that he falls off his horse (as Silenus does immediately after the quoted lines in the *Ars amatoria*; cf. 1.547), and by the fact that, shortly after the Cook has been helped back onto his steed, the Host praises

introduced the disruptive Miller as, at least in part, a figure of Bacchic energy and non-constraint whose intervention frees the storytelling from the straitened notions of decorum which the Host seems intent to impose upon it by calling on the pilgrims in a descending order of social rank. Meanwhile, at another level, the Miller serves as a comic metaphor for Chaucer's inspiration to transform his narrative by introducing into it the notion of rivalry — between pilgrims (the Miller and the Host, and even more the Miller and the Reeve), and between stories (the Miller "quites" the Knight's Tale by ridiculing its rhetorical and philosophical implications in his own relentlessly concrete and reductive storytelling) — that uniquely enlivens the framing fiction of the itinerant "compaignye of sondry folk" (GenPro 24–25).

But there is still another dimension to the Miller that Chaucer evokes by the vivid, Ovid-inspired image of him barely astride his horse. This is, after all, a time-honored metaphor for lost control;[11] in fact, toward the end of the *Canterbury Tales,* another drunken figure, the Cook, does actually fall off his horse, and is only restored to it with effort (MancProl 46–55). Since the pilgrims, according to the framing fiction, are riding on their way to Canterbury to offer thanks or seek favors at the shrine of the "hooly blisful martir" (GenPro 17), inability to stay astride one's horse, and therefore to keep a steady course for one's goal, carries with it suggestions of moral errancy and lost initiative — a truancy from Christian virtue suggested by the memorable phrase, "wandrynge by the weye," which Chaucer uses (GenPro 467) in connection with the Wife of Bath, another character who challenges decorum and establishment mores. The Miller's posture, both physical and verbal, is one of waywardness as he challenges the Host's authority — "For I wol speke or elles go my wey" (MilProl 3133) — and the normative virtues of the Knight's "noble storie" (3111): "By armes and by blood and bones,/ I kan a *noble tale* for the nones,/ With which I wol now quite the Knyghtes Tale" (3125–27; emphasis added).

The Miller's subversion of the judgment of "gentils" and Host — i.e., of the order established in society at large and in this itinerant microcosm of it — puts him in the role of rebel, and the Host's disgusted words to the drunken Robyn, "Tel on, a devele wey!" (3134) remind us of the archetypal rebel under whose colors the Miller has in effect

Bacchus (MancProl 99–101 — the only place in the *Tales* where this god is specifically mentioned).

11. See A. Bartlett Giamatti, "Headlong Horses, Headless Horsemen: An Essay on the Chivalric Epics of Pulci, Boiardo, and Ariosto," in G. Rimanelli and K. Atchity, eds., *Italian Literature: Roots and Branches* (New Haven, 1976), 265–307, for a brief history of the metaphor in European epic literature.

placed himself. According to medieval Christian belief, the origin of evil in the universe was the rebellion of Lucifer against God, an act of pride that took the form of a challenge to the divine authority and kingship. Lucifer's presumption (dramatized in the medieval English mystery plays as his insistence on seating himself on God's throne), was punished by his fall, with those angels who followed him in his rebellion, from that place of eminence into the newly created Hell, and by his transformation from the brightest angel into the devil. As Robyn begins his rejection of the Host's authority (to which he, along with the rest of the pilgrims, has earlier committed himself) (cf. GenPro 810–18), his near fall from his mount almost reenacts, at the level of low comedy, the consequence of the universe's first such confrontation. Chaucer, I suspect, wants us to appreciate not only the precariousness of Miller's control over his horse, but also the precariousness of the Host's control over his flock and enterprise, and indeed the ease with which the ad hoc pilgrim society can, through the Miller's subversive tale telling, lose its coherence (established by communication and "forward" at the Tabard), fall into a state of ill-natured quarreling, and perhaps fall apart. (The Miller, after all, threatens to leave the company if he does not get his way.) Here would be "a devele wey" indeed.

In short, the Miller's injection of Bacchic energies into the storytelling is also his (and Chaucer's) deflection of the pilgrimage fiction into the realm of fallen language: the language of rage, sarcasm, sensual license, and even blasphemy.[12] That such discourse fascinates us is borne out by the perpetual popularity of fabliaux and analogous stories of ribaldry and uproar; that it fascinated Chaucer would seem to follow from the tremendous artistry with which he invested those of *The Canterbury Tales* (such as the Miller's, the Merchant's, and the Summoner's), most closely related to the fabliau. Nonetheless, as the apology for telling "cherles tale[s]" (MilProl 3162–86; cf. GenPro 725–42) that Chaucer puts in the narrator's mouth suggests, the poet was aware that language embodying the vices of fallen humanity opened itself to condemnation on religious or moral grounds. Indeed, his Parson articulates a Puritan position toward fiction, contrasting "fables

12. See, respectively, MilProl 3143–49 (the Reeve's angry interruption of the Miller); 3151–57 (featuring the Miller's cynical maxim, "Who hath no wyf, he is no cokewold" (3152); 3163–66 (the Miller's insistence that a husband should not worry about his wife's fidelity as long as there is enough sex left for himself); 3141–43, 3163–64 (the Miller's description of his fabliau in words designed to recall the Infancy narrative of the Gospels, and his equation of "Goddes pryvetee" with a woman's sexuality).

and swich wrecchednesse" that "weyven soothfastnesse" with the "moralitee and vertuous mateere" he proposes to recount in his (prose) treatise (ParsProl 31–47). So the Miller's Ovid-derived state of poor balance becomes, in Chaucer's hands, a usefully ambivalent emblem of exhilarating release from artistic restraint and dangerous descent into disreputable discourse — an emblem, that is, of *The Canterbury Tales* as a poetic project.[13]

The other allusion in the "Miller's Tale" that shows Chaucer's imagination characteristically engaged in appropriating and transforming "classical" lore to his own narrative (and comic) purposes comes within the Tale itself, at a moment early in an Oxford student's plot to enjoy the sexual favors of a beautiful young townswoman. Nicholas, brimming over with youthful self-assurance and clerical snobbery, is convinced that "a clerk hadde litherly biset his whyle,/ But if he koude a carpenter bigyle" (MilT 3299–3300), and goes about concocting a ridiculous (and ridiculously complicated) scheme whereby he will cuckold the particular carpenter, John, from whom he rents a room and on whose wife, Alisoun, he has lecherous designs.

Nicholas undertakes to convince John that a second flood, greater than Noah's, is about to engulf the world. To establish his credibility as a seer, our ingenious hero locks himself in his chamber and pretends to be sunk in a trance, "capyng evere uprighte,/ As he had kiked on the newe moone" (3444–45). When John discovers this state of affairs, he is greatly upset at the thought that Nicholas has been driven mad by delving into the mysteries of the heavens (we have earlier been told of Nicholas that "al his fantasye/ Was turned for to lerne astrologye" [3191–92]):

> "This man is falle, with his astromye [sic],
> In som woodnesse or in som agonye.
> I thoghte ay wel how that it sholde be —
> Men sholde not knowe of Goddes pryvetee.
> Ye, blessed be alwey a lewed man
> That noght but oonly his bileve kan!
> So ferde another clerk with astromye;

13. See, for example, Derek Pearsall, *The Canterbury Tales,* Unwin Critical Library (London, 1985), 29–48, on Chaucer's adaptation of the literary form of the frame-tale collection in order to obtain "freedom from a premature and restrictive commitment to fixed attitudes or interpretations" (45); and Alfred David, *The Strumpet Muse: Art and Morals in Chaucer's Poetry* (Bloomington and London, 1976), especially 52 ff., on how, "through the invention of the frame story, Chaucer is able to escape his moral obligations as a poet" (75).

He walked in the feeldes for to prye
Upon the sterres, what ther sholde bifalle,
Til he was in a marle-put yfalle —
He saugh nat that!" (3451–61)

Alistair Minnis contends that John's reference to the clerk who falls into a fertilizer pit while seeking to read the future in the stars is an allusion to the popular medieval story of the emperor Alexander and his tutor, the astrologer-magician Nectanabus.[14] (John Gower, Chaucer's contemporary and fellow poet, retells the story in his *Confessio amantis* [vi.1789–2366].) To Chaucer, this story would presumably have seemed as authentically "ancient" (and authoritative) as any tale from Ovid or Virgil. As Minnis recapitulates, "Nectanabus, having appeared in the likeness of the god Jupiter, corrupts and has his pleasure of Olympia, Queen of Macedonia. Alexander is begotten as a result of this incredible encounter....When Nectanabus is teaching the young Alexander astrology, his cynical pupil pushes him into a deep pit. Alexander proceeds to marvel at the ignorance of his tutor, who ought to have been forewarned of this event by the stars. Nectanabus replies that no man can flee from his fate: he knew by his science that his own son would cause him death."[15] Thereupon the dying sage reveals his paternity to the surprised monarch; thus "Alexander's very attempt to shrug off fatalism has confirmed Nectanabus's prophecy and sealed his fate."[16]

Medieval clerics, such as Vincent of Beauvais and Ranulph Higden, who included the story of Alexander and Nectanabus in Latin works of encyclopedic and didactic intent "portray these pagans as fatalists who cannot escape their fate."[17] Alexander's act in killing Nectanabus only proves what he seeks to disprove: that the stars control our destinies. The Herodotean irony and pessimism of the story insured its fascination for generations of readers and listeners, but also enshrined a fatalism that "would...seem to be a viable philosophy for a pagan — but not for a Christian"[18] since, as John of Trevisa

14. *Chaucer and Pagan Antiquity,* 63–64. Chaucer is doubtless alluding as well to the story about the philosopher-astronomer Thales of Miletus, who fell in a ditch while looking at the stars. For medieval versions of this story, see D. W. Robertson, *A Preface to Chaucer* (Princeton, 1962), 273.
15. Page 63.
16. Ibid., 64.
17. Ibid., 63.
18. Ibid., 64.

points out in a comment on the tale, "from every mishap that man is i-schape in this worlde to falle ynne, God may hym save if it is his wille."[19]

Another feature of the Alexander-Nectanabus story that undoubtedly contributed to its popularity, and has relevance for Chaucer's allusion to it, is its portrayal of father-son relations, a central issue in any society as patriarchal as medieval Europe. Here the son only learns his father's identity in the process of destroying him; the full realization of the relationship thus coincides with its definitive rupture. A further bitter paradox concerns the circumstances of the father's death, which comes as he teaches his son a profound truth (that all are fated and can read their fates in the stars). The son, rebelling against the father's tuition, attempts to disprove it, and in doing so provides a pefect demonstration of its accuracy. But the act which vindicates the teacher simultaneously destroys the father. The link between this story and the Oedipus myth — with pedagogy substituted for politics — is chillingly obvious.

By contrast with this dark vision, John's exclamation outside Nicholas's locked door is a masterpiece of comic diction, narration, and characterization. First of all, the tragedy of the Alexander-Nectanabus story suffers a radical deflation to bathos. The "clerk" whom the carpenter sees as a parallel to, and an *exemplum* for, Nicholas ends up in a pit not as the result of a doomed attempt to disprove fate, but because he wasn't watching where he was going. The fall proves not that humanity is trapped in an irrevocable destiny, but that eggheads don't have the common sense to keep their eyes on the ground in front of them because they're too busy "prying" into "Goddes pryvetee," which isn't their business in any case. Indeed, the worried carpenter presents the vignette (for such it has become in his mouth) in support of the notion that the ignorance of the "lewed man," who can only repeat credal formulae, is much to be preferred to intellectual meddling and "astromye" that court the ultimate bathetic disaster: the pratfall. Alexander's unwitting patricide has been metamorphosed into the medieval equivalent of a skid on a banana peel; the marl-pit, with its suggestion of manure,[20] provides an occasion for the ultimate disgrace of the haughty cleric and a triumphant vindication of the common folk whom he habitually scorns, but who can now, with the tables turned,

19. Quoted by Minnis, 64.

20. See the *Oxford English Dictionary* (Oxford, repr. 1971), entry "marl" vb 1: 2, "to enrich, as with marl [a kind of clay], to manure, to fertilize." The earliest citation for this meaning is dated 1544. Nonetheless, the satiric intent of John's story would be best served, and its protagonist's plight all the more embarrassing, if the term contained an excremental implication.

laugh at him ("He saugh not that!"). Pagan tragedy has become fabliau revenge-farce.

This understanding of John's vestpocket *exemplum vitii* makes of it an apt, if unwitting, reply to Nicholas's breezy assumption, quoted above, that any clerk worth his salt can easily get the better of a mere carpenter. But in his comic allusion to the Alexander-Nectanabus story, Chaucer also revises the portrait it presents of father-son relationships. No sooner has John uttered his gleeful comment on the star-gazer who couldn't see the pit at his feet, than his tone changes markedly as he returns to the case at hand, the entranced Nicholas: "But yet, by Seint Thomas,/ Me reweth soore of hende Nicholas./ He shal be rated of his studying..." (3461–63). This is the voice of the worried father, promising to scold his son for reading in bad light, not the blue-collar worker's cry of triumph at the discomfiture of the know-it-all professor. In short, it is the son who has "fallen" (into a pit [3460], into "woodnesse" [3452]) and the father who, far from having pushed him in, will now try hard to extricate him.

Even as he reverses or deflates the tone of his inherited story, however, Chaucer also retains, in suitably transformed guise, some of the irony that marks the Alexander-Nectanabus tale. Although the old carpenter adduces the story of clerical comedownance as a metaphor for the pitfalls of the "astromye" by which his young boarder has been seduced, the real trap is being laid by the "son," to get the "father" out of the way so that he can be cuckolded. John's gullibility leads him to cooperate in digging his own pit, or, to use the tale's concretization of the metaphor, in hanging his own tub. There he will sleep (albeit uncomfortably) while Nicholas usurps his bed, until the clerk's anguished cry for water leads to John's literal fall from under his roof (where he has hung the tub, at Nicholas's direction). The physical injury he sustains — a broken arm — recalls, though it does not equal, Nectanabus's fate. Moreover, the carpenter's earlier conviction that Nicholas "is falle in som woodnesse" now rebounds on him: when he tries to explain his hurtful descent to the neighbors who have "in ronnen for to gauren on this man,... it was for noght, no man his reson herde./ With othes grete he was so sworn adoun/ That he was holde wood in al the toun" (3827, 3844–46).

One further function of Chaucer's allusion to the Alexander-Nectanabus story remains to be noted: its contribution to a comic consideration, within the Miller's Tale, of Christian attitudes toward questions of fate and providence. Nectanabus' death, we recall, proves both the inexorability and the predictability of the astrologer-magician's fate. Late medieval orthodox Christianity, in avowing God's control over all creation (including humanity), vigorously denied the absolute power of the stars over human destiny and therefore the absolute (though often not the partial) efficacy of astrology. John's bathetic version of the "classical" story omits Alexander, whose assault on his father illustrates the irony of accomplishing fate while seeking to avoid or deny it, and in effect transforms Nectanabus into a kind of absent-minded professor whose fall constitutes a wonderfully concrete (not to say reductive) paradigm of both the dangers of astrology and the fact of personal responsibility. John's pious ejaculation, "Help us, Seinte Frydeswyde!/ A man wot litel what hym shal bityde" (3449–50), establishes him, as it were, as a Christian anti-Nectanabus. Yet, within moments of his assertion that "Men sholde nat knowe of Goddes pryvetee" (3454), John has swung about to become an anti-Alexander, that is, a complete believer in the truth of what his "tutor," Nicholas claims to have discovered about the divine plan "in myn astrologye" (3514).

In Chaucer's hands, the ironic pessimism of the Alexander-Nectanabus story is reworked into the stuff of parody and burlesque. His genius for appropriating and transforming "classical" material shines through even the brief, sometimes minimal allusions that have come under scrutiny in this essay. Chaucer rejoiced in his mastery of inherited literary and cultural traditions. He was indeed a giant standing on the shoulders of giants.

■

X
Donatello and the Brancacci Tomb in Naples

JAMES BECK

THE MAGISTRAL SEPULCHRAL MONUMENT for Cardinal Rinaldo (sometimes written as Rainaldo) Brancacci now in the right transept of S. Angelo a Nilo in Naples is a widely recognized collaborative enterprise produced in the Early Renaissance by Donatello and his partner of the time, Michelozzo di Bartolomeo, along with their assistants [Fig. 1]. In the Florentine Catasto of July 1427, these artists declare having received 300 out of a total of 850 florins promised for the work. To be sure, it was only mentioned in Michelozzo's *portata*, where it is clearly indicated as belonging to the partnership. In addition, Michelozzo also filled out Donatello's report so there could hardly have been any attempt to hide the facts.[1] Following the Council of Constance, Rinaldo had been in Florence for an extended stay when the papal court moved there, before, that is, the definitive return to Rome.[2] The cardinal died in Rome on 5 June 1427 (not, as frequently reported, on 27 March 1427, which is the date of his will).

1. For a good summary of the documentary evidence regarding the monument see H. W. Janson, *The Sculpture of Donatello* (Princeton, 1963), 88–90 and H. McNeil Caplow, *Michelozzo* (New York, 1977), 1:154–63.
2. A family connection between the Neapolitan Brancacci and the Florentine family of the same name is engaging from an art historical point of view because of the potential interrelationships of patrons of Donatello and Masaccio, who were themselves intimate friends. It is widely asserted that there was no relation whatsoever (among others, see Janson, *Donatello*, 89, n. 1) between the Neapolitan and Florentine families. On Rinaldo, see R. W. Lightbown, *Donatello and Michelozzo: An Artistic Partnership and its Patrons in the Early Renaissance* (London, 1980), 1:52–82, who insists on calling the family name Brancaccio. On the other hand, it is known that Rinaldo stayed in the palace of the Florentine Brancacci when the court was in Florence, as pointed out in J. Beck, *Masaccio: The Documents* (Locust Valley, NY, 1978), 55. The information comes from the fascinating "Diario fiorentino di Bartolommeo di Michele del Corazza, anni 1405–1435," in *Archivio storico italiano*, ser. 5, 14 (1894). This reference did not escape Caplow, *Michelozzo* 1:142n; and also Lightbown, *Donatello and Michelozzo,* 1:62 and 2:265, n. 27.

H.W. Janson has pointed out in his fundamental study on Donatello that the tradition of Cosimo's presumed role as patron of the Brancacci Tomb cannot be documented, but he was nonetheless inclined to give it "some credence." H. McN. Caplow, following Janson, found that the first time the assertion was made that Cosimo had ordered the tomb for Cardinal Brancacci was in the seventeenth century, fully two centuries after the fact. She concluded that the claim was only "partially true."[3]

Undoubted there are connections between the commissioning as well as the execution of the Tomb of the Neapolitan antipope (as John XXIII) Baldassarre Coscia and that of his trusted relation Rinaldo Brancacci. The Coscia monument was placed in the Florentine Baptistery, though only after considerable negotiations on the part of the executors. A Medici component was present in both. Giovanni di Bicci, father of Cosimo and Lorenzo, was an executor of John XXIII's tomb, a posthumous one, assigned to the partners (or, to Donatello alone, at the beginning) subsequent to the death of the recipient. The Brancacci monument commission was given out during his lifetime, although probably not directly to the sculptors, but through an intermediary. While it is quite certain that work on the Brancacci sculpture was done in Pisa with the specific intention of sending the finished sculptures down to Naples by ship, the precise location of the carving on the John XXIII monument is less securely determined. The bronze effigy was almost certainly produced in Florence, for there would have been no purpose whatsoever in doing it in Pisa, while the marble sculpture could have been carved either in Pisa or in Florence. To my knowledge, no one has raised the possibility that it might have been done, even in part, in Pisa. On the other hand, it is likely that while the *bottega* was working on the marble statues and the reliefs, they also were doing the marble work for the Florentine assignment in the same Pisan shop.

As for the time when the sculpture was being produced, in the most general terms, it can safely be assumed that work on both the tombs was executed during the second half of the decade 1420–1430. Lightbown dates the Brancacci monument to c. 1426–1428, for example.

Payments emanating from the bank of Averardo di Francesco de' Medici and dating to the year 1426 which were made to

3. Caplow, *Michelozzo*, 1:156–57; and C. d'E. Caracciolo, *Napoli sacra* (Naples, 1624), 261.

Donatello alone, that is, without mention of his partner Michelozzo, are considered by Janson as being connected with the Brancacci Tomb, at least in part.[4] Since both the Brancacci Tomb and the Coscia Tomb were joint ventures, basic questions remained unanswered. Perhaps these sums were made out to Donatello for other works, even ones for the Florentine Medici: for example, for their palace or for the sacristy of San Lorenzo then being constructed? On the other hand, we cannot exclude the possibility that the payments were for one or both of the monumental tombs being worked.

The payments were made in Pisa, to be sure, but the funds originated in the Florentine bank of the Medici, namely that of Cosimo and Lorenzo di Giovanni. Furthermore, the piecemeal payments do not seem to have the flavor of referring to large-scale purchases of stone that would have been necessary for the grandiose, monumental tomb projects. On the other hand, the total amount paid in Pisa in 1426, really in a matter of a few months, was slightly over 120 Florins, a tidy sum. When added to the 188 Florins paid to or at least owed to the partners during the following year, the total of 308 Florins is suspiciously close to the advances that the two (really Michelozzo) claimed to have received up to the time of the Catasto of 1427.[5]

4. These documents were first published (incompletely) by D. Semper, *Donatello, seine Zeit und Schule* (Vienna, 1875), 280 and 310. They are noted by Caplow, *Michelozzo,* 1:158; Lightbown, *Donatello and Michelozzo,* 1:78–79, but were ignored by V. Herzner, "Registi donatelliani," *Rivista dell'Istituto Nazionale d'Archeologia e Storia dell'Arte,* ser. 3, 2 (1979): 169–228. B.A. Bennett and D.G. Wilkins, *Donatello* (Oxford, 1984), took into consideration these payments for the Brancacci Tomb and assumed that Donatello was in charge of the project and consequently the payments were made to him alone.

Parenthetically one should commend Herzner for his highly serviceable compendium of documents, but it in no way precludes the publication of the complete documentation on Donatello, as has been done with Raphael by V. Golzio, *Raffaello nei documenti, nelle testimonianze dei contemporanei e nella letteratura del suo secolo* (Vatican City, 1936); and the new, vastly enlarged edition by J. Shearman, which is in press.

5. The 188 Florins occurs in the portata of Cosimo and Lorenzo de' Medici and may be found in ASF, Catasto, no. 5l, fols. 1162–1168v. See Caplow, *Michelozzo,* 1:156, n. 28.

The *portata* of Michelozzo is most recently published in M. Ferrara and F. Quinterio, *Michelozzo di Bartolomeo* (Florence, 1984), 37–38. In it he spells out his partnership with Donatello and mentions the projects upon which they were then, that is at the time of the report in 1427, collaborating. They were the Tomb of John XXIII, that of Rinaldo Brancacci, the Aragazzi Tomb monument, and a large statue for the Duomo, not securely identified but probably the *Zuccone*. For these works, a total of 1037 Florins had already been received by the partners, of which Michelozzo claims that his share was 200 Florins.

Considering the inherent ambiguities and difficulties of interpretation, perhaps it is best to lay out the evidence.[6] Below are the payments to Donatello found by the bank of Averardo di Francesco de' Medici and which are located in the Archivio di Stato, Firenze (ASF), MAP, No. 3.[7]

[fol. 104v modern, CVv originally]

+ MCCCCXXV

[1] Donato di Nicholo intagliatore de' dare adì vi d'aprile [1426] fiorini due d'oro, F. ii

[2] E adì viii d'aprile fiorini tre d'oro per lui a Piero di Bertino, fiorino 2 portò contanti, e resto portò e'detto in grossi

F. iii

[3] E adì detto grossi venzette, per lui a Lodovicho Tanaglia, portò e' detto e sono per chosto di 2 paia di chalze

F. i s. xvii d.ii

(This sum is confirmed in the more summary form as it appears in the *campione*, for which also see Ferrero and Quinterio, *Michelozzo*, 39–40). This is a highly informative but entirely neglected admission, and should make it quite clear that Donatello had a leading role in the commissions up to the moment of the declaration. For the Aragazzi monument for Montepulciano the advance given up to the time of the Catasto was 100 Florins, and that for the marble. We may assume that Michelozzo had a much greater role in its production than in those of the others.

6. R. de Roover, *The Rise and Decline of the Medici Bank, 1397–1494* (Cambridge, 1963), 421, n. 46, assumed that it was Cosimo who paid Donatello these advances through the Pisa branch of the Medici bank, when referring to these payments. That author also made the interesting observation that Donatello belonged to one of the the poorer branches of the Bardi. Keep in mind that at this time Cosimo and Lorenzo were in partnership with Ilarione de' Bardi in Florence and that Cosimo was married to Contessina de' Bardi.

7. Averardo di Francesco de' Medici was a cousin of Cosimo and Lorenzo, who had no branch of their own in Pisa. Averardo was a man of great power and distinction and was one of the Sei della Mercanzia, serving along with Niccolò da Uzzano from January to June 1431. There are several letters written by Michelozzo to Averardo of 1430 (see Ferrara and Quinterio, *Michelozzo*, 33–34). For Averardo see D. Kent, *The Rise of the Medici Faction in Florence, 1426–1434* (Oxford, 1978), passim.

In an original *ricordanza* written by Francesco di Giuliano d'Averardo de' Medici of his marriage with Costanza di Piero Guicciardini, a fascinating document, there is the following (ASF, MAP, No. 148, pezzo 31, fol. 3): "...e faciemo un aparato più bello che già molti anni si fusse facto, acchonciolo Michelozo intagliatore...." The event took place on 14 June 1433 and offers still another facet of Michelozzo's remarkable career which also included sculptor, architect, goldsmith, and diemaker. The marriage in mentioned by Kent, ibid., 45, 46 and 56, n. 18, including a reference to the document. It is worthy of a complete publication.

[4] E adì xiii d'aprile fiorini uno d'oro, portò detto in grossi
F. i

[5] E adì xvi d'aprile fiorini dieci d'oro, per lui a Lorenzo di Pellegrino da Torano, portò e' detto chontanti e sono per parte del marmo di più lavori à fatto
F. x

[6] E dì xx d'aprile fiorini due d'oro, portò e'detto in grossi
F. ii

[7] E adì xxx d'aprile fiorini tre d'oro, portò e'detto in grossi, e qua' dise voleva fiorini 2 per 1[n]o pezo di maro, e resto per ispese
F. iii

[8] E adì xi di magio fiorini uno d'oro, portò e'detto in grossi
F. i

[9] E de'dare adì xi di magio fiorini vi soldi ii denari x a oro, posto debbi aver in questo a c. 113, per resto di questa ragione

[total] 30.0.0 F vi s. ii d.x

fol. 105 [modern, CVI, originally]

+MCCCCXXV

Donato di Niccholò intagliatore de'avere adì vi d'aprile fiorini trenta d'oro, per lui da' Medici di Firenze, mes[s]i a uscita segnata H, c. 121
F. xxx

fol. 111v [modern, CXIIv, originally]

+MCCCCXXVI

[1] Donato di Niccholò intagliatore de'dare adì 17 di magio fiorini due d'oro, portò e'detto in grossi
F. ii

[2] E adì xxiiii di magio fiorini uno d'oro, portò e'detto in grossi
F. i

[3] E adì 29 di magio fiorino uno d'oro, portò e'detto in grossi
F. i

[4] E adì viii di giugnio fiorino uno d'oro, portò e'detto in grossi
F. i

[5] E adì xii di giugnio fiorino uno, portò e' detto contanti
F. i

[6] E adì xiii di giugnio fiorino uno d'oro, portò in grossi
F. i

[7] E adì xx di giugnio fiorino uno d'oro, portò e'detto in grossi
F. i

[8] E adì detto fiorini sei d'oro, per lui a Ser Andrea di ser Rufino, fattore dell'Opera, portò contanti e sono per la [=una] prieta comperò dall'Opera F. vi

[9] E adì xxvi di giugnio fiorini tre d'oro, portò e'detto in grossi F. iii

[10] E adì xxx di giugnio fiorini quatro d'oro, portò e' detto in grossi F. iiii

[11] E adì svi di luglio fiorini quatro d'oro, per lui a 'Ndrea di Vita da Trapani, portò e'detto in grossi F. iiii

[12] E adì vi d'agosto fiorini uno d'oro, portò e'detto in grossi F. i

[13] E insino adì iii d'aghosto fiorini sedici d'oro, per lui a Piero Chaetano, portò...[blank in text] e sono per la [=una] promessa io Luigi fe' a detto Piero per la barcha comprò detto Donato F. xvi

[14] resto [in margin] E de'dare fiorini quarantacinque soldi ii denari x, posto deba avere in questo a c. 127 [missing] F. xlv s. ii d. x

[total] 90.2.10

[fol. 112 modern, CXIII, originally]

+MCCCCXXVI

[1] Donato di Niccholò intagliatore de'avere adì xi di magio fiorini sei soldi ii denari x a oro, per resto di suo ragione a dietro, posto debbi dare in questo a c. 106 F. vi s. ii d.x

[2] E adì 18 di giugno, fiorini quatgro d'oro, per lui da Sarafino di Lorenzo del Biada e per lui da Giovani Cirioni rechò Luigi d'Astore in grossi F. iiii

10.2.10 [between the lines, partial total]

[3] E adì vi d'aghosto fiorini ottanta, per lui da' Medici di Firenze mes[s]i a uscita segnata H a c. 128, e sono per la lettera di chanbio F. lxxx

90.2.10 [total]

Some help in straightening out the situation with regard to the Brancacci monument can be supplied by several hitherto completely unknown documents found in an account book of the bank of Cosimo and Lorenzo de' Medici and Ilarione de' Bardi. Although there appears to be a discrepancy in the number of *fiorini di camera*, either 850 or 950, the sum calculated in *fiorini fiorentini* remains constant at 900, and that seems to have been the final price of the tomb. It corresponds, although not exactly, to the 850 florins claimed to have been the agreed-upon sum by the artists in 1427. These documents, like the previous ones, are located in ASF, MAP, no. 153, ii ["Libro secreto"][8]:

[fol. 26]

[Messer Rinaldo Cardinale de' Brancacci] E dì xxiiii di marzo 1428 [= 1429, modern] fiorini ottocento cinquanta di camera, per lui faciemo buoni a'nostri di Firenze in fiorini VIIII C [above the line] che chosì li ragionamo, posto li debino avere in questo a c. 47, sono per fiorini DCCCL di camera si à pagare per la sua sipultura, chome ordinorno Cosimo de' Medici e C[ompagn]i a Donato di Niccholò e Michelozo di Bartolomeo intagliatori, che ànno tolto a fare detta sipoltura

F. DCCCC

[fol. 47]

[Chosimo e Lorenzo de' Medici e compagni di Firenze dono avere] E dì detto [24 Marzo 1428 = 1429, modern] fiorini novecento d'oro per fiorini DCCCCL di camera faciemo loro buoni per messer Rinaldo Cardinale de' Branchacci, posto li debi dare in questo a c. 26, sono per la sua sipoltura fa fare Cosimo nostro F. DCCCC

As is often the case with accounting documents of this kind, the questions that particularly interest modern historians are not the ones that motivated the bank clerks. Nonetheless we appear to have for the first time confirmation that Cosimo de' Medici had indeed commissioned Donatello and Michelozzo to produce the Brancacci monument, as well as seeing to the payments. Here then is evidence helpful in developing a profile of Cosimo as a patron of the visual

8. For the documentary context of these items, see Appendix I.

arts. In this regard, he is usually given a secondary role, with his son Piero the Gouty thought to have been the expert for such matters. There is growing evidence that Cosimo was very much devoted to the potentials of sculptural decoration as is evidenced by the elaborate program of the Sacristia Vecchia, which was apparently his project, one that was to commemorate his father (and mother) as well as the Medici family in general.[9]

Is it possible from these two entries to achieve the date of 24 March 1429 as that of the completion of the Brancacci Tomb? While they serve to confirm the final price, it is still not altogether clear whether the payments were merely to debit the account of the cardinal, passing it to the bank, only later to be distributed to the sculptors, or not. Since the transaction took place on the last day of the Florentine year, with the new year, as is well known, beginning on 25 March, it may have been made for purely accounting reasons. On the other hand, since the sculptors are specifically mentioned, it would seem that the money had either been given to them already, or partially paid with a portion still outstanding. With some caution, I suggest that this date can be taken as the *terminus ante quem* for the sculpture and its shipment to Naples. Precisely when the monument was actually set up in S. Angelo a Nilo, is another matter, but Janson made the point that in the tax returns of 1430 and thereafter the monument is never mentioned again, a fact which led the author to think that it was already finished by that time. The same writer has also pointed out that Donatello and Michelozzo were apparently in Pisa in mid-1428, and had been there for some time, making it likely that they had finished their work there by the end of

9. In what must be thought of as a first-hand and authorative account, Vespasiano da Bistici in his life of Cosimo, ed. A. Greco, *Le Vite* (Florence, 1976), 2:193, leaves us with the following account:

> [Cosimo] Era tanto universale in ogni cosa, che con tutti quegli parlava aveva materia. Era tanto universale in ogni cosa che tutti quegli che andavano a parlare co'lui, secondo le loro facultà, cor ognuno sapeva essere et con tutti aveva materia, s'egli era cor uno literato regionava della sua facultà...se praticava con pittori o scultori egli se ne intendeva assai, et aveva alcune cose in case di mano di singulari maestri. Se di scultura, egli n'era intendentissimo, et molto favoriva gli scultori et tutti artifici degni. Fu molto amico di Donatello et di tutti e' pittori e scultori....

Vespasiano's statement seems to disprove E.H. Gombrich's claim, *Norm and Form: Studies in the Art of the Renaissance* (London, 1966), 36, that when "the Medici first appear in their role as patrons, their activity still fits completely into the age-old traditions of religious life."

1428, when he dates the completion of the work and the shipment of marbles to Naples.[10]

Although the patronage question and the dating of the Brancacci Tomb has been tidied up, if only a bit, by the documents presented here, the vexing issue of the division of labor on the tomb still requires a comment. Recent criticism, as exemplified by Janson, has been reluctant to give Donatello anything but a minor share on the project.[11] If we consider, however, the importance of the patron (Cosimo) as well as his unique relationship to Donatello which was to continue until his death in 1464, as well as the admittedly minor share that Michelozzo held, at least up to the middle of 1427 (less than 20 percent), Donatello's presence must have been much more significant than has been assumed: and this fact is true not only for the Brancacci monument but also for the John XXIII Tomb.[12] There is universal agreement only that Donatello produced the *Assumption* relief [Fig. 1], and for most interpreters, his intervention stopped with this somewhat minor plaque.[13]

This is not the occasion to give still another detailed visual analysis of the sculpture on the tomb, especially in light of the recent work of Janson, Caplow, and Lightbown. On the other hand it is useful to consider that there are a number of highly innovative features to the monument, that certain passages of the sculpture reveal high invention, and that Donatello at least on the early stages had by far the greater share in producing the work. With this in mind, it is perhaps best to see a more significant contribution on Donatello's part than is now assumed. Because of the virtually inaccessibility of the monument especially in the more recent past, the unavailability of high-quality photographs, and the difficulty in viewing directly most portions of the tallish monument, the normal procedures for first-hand examination and comparison with other works by Donatello and Michelozzo is particularly difficult. Indeed only in 1958 was it fully realized that most of the sculpture is unfinished to one degree or another,

10. Janson, *Donatello*, 90, n. 7.
11. Ibid., 90.
12. For the Aragazzi monument we know that at the start the advance was 100 florins that was was presumably credited equally to the two partners, although when work actual proceeded, Michelozzo must have had the lion's share.
13. See above, note 5. In her carefully observed and reasoned analysis of the various portions of the Brancacci monument, Caplow nonetheless remains among those modern scholars who find only a very slight share for Donatello. Vasari and older commentaries give Donatello the major role.

making effective attributions between Donatello, Michelozzo, and shop assistants all the more difficult.[14]

If there are good chronological and documentary reasons why this project should not be so readily dismissed for Donatello, creative features of the monument also support a Donatellean role. Habitually critics have tended to award to Michelozzo all the architectural projects, or architectural aspects of projects, that the partnership had undertaken, simply because Michelozzo became a distinguished independent architect later on. During the 1420s there is no evidence of architectural activity on his part, and besides, the Tabernacle Tomb for Rinaldo Brancacci seems to be more of a sculptural solution than an architectonic one. Even if some of the details appear to be old-fashioned for the progressive Donato, especially the crowning element with its "gothic features," there is still insufficient reason to rule out Donatello. In order to harmonize the monument within the Neapolitan environment, a trecento, that is to say traditional, flavor must have been agreed upon between Cosimo, relatives of the deceased, and the sculptors. In any case Donatello had easy access to his friend Brunelleschi for advice, were that necessary, and in the case of the innovative Mercanzia niche for St. Louis of Toulouse, one is inclined to see in it a contribution by Brunelleschi rather than Michelozzo. Furthermore it has been widely observed that there are connections between the *Trinity* by Masaccio and the Brancacci Tomb: the unifying design element is not Michelozzo but Pippo. Nor is there any reason to believe that Donatello did not feel confident when it came to designing such objects, notwithstanding Brunelleschi's negative evaluation of his architectural skills in the Sacristia Vecchia.

Perhaps the most striking insight in the development of the new style was the Caryatid-Virtues which support the sarcophagus. They are, to be sure, a trecento survival with examples from all over Italy, including Naples, but among the most interesting are those in the Duomo of Pisa by Giovanni Pisano. Of course, they represent a classical revival, or at least reinterpretation of classical forms. But most importantly, as conceived in the Brancacci and Aragazzi Tombs, and unlike medieval prototypes, they belong to a category of independent life-size marble statues conducted in-the-round, the first in the Renaissance.

14. V. Martinelli, "Il non-finito di Donatello," *Donatello e il suo tempo: Atti dell' VIII convegno internazionale di studi sul rinascimento, 1966* (Florence, 1968), 179–94; and Caplow, *Michelozzo,* 1:180–209; Lightbown, *Michelozzo and Donatello,* 1:121 ff.

In considering attributions of various portions of larger, complex works such as this one, the general operative rule is that everything that is "supreme," is by the most renowned master, in this case, Donatello. Then, the best of the second-level works must be by Michelozzo, and and truly inferior portions are by modest shop assistants. Some scholars have attributed one or more of the Caryatids to Donatello or Donatello's shop (somehow, to me, a studied avoidance of the issue), as opposed to Michelozzo or his assistants, but without much conviction. The Caryatids certainly differ among themselves in conception and style, with the side two being flatter and more mannered than the central image. Following a recent first-hand examination, I am convinced that the central figure is by Donatello.[15] It differs from the other two in having been conceived more fully as a statue in-the-round, and without illusionism. It is also severely classical and monumental, while the head is *all'antica*[16] [Fig. 2]. The heavy swag of drapery that covers the front of the figure is connected with the *St. Louis of Toulouse* and the *Zuccone*, perhaps falling chronologically between them [Fig. 3].

In robust conception the central Virtue, which like the other two has no attribute, is related to the two large nude trumpet-blowing putti on the uppermost zone of the tabernacle [Fig. 4]. They are remarkably precocious figures, also conceived as in-the-round elements, and are expanded cousins of the bronze statuettes Donatello produced for the Baptismal Font in Siena at about the same time. That they are indeed more restrained in movement than the bronzes from Siena, as observed by Caplow, is explained by their function and by the restraints of the material, in this case, marble as opposed to the more flexible cast bronze.[17]

I have much less difficulty in giving Donatello a share, and a determining one at that, in the effigy figure of the deceased cardinal,

15. See J. Pope-Hennessy, *Italian Renaissance Sculpture* (London, 1958), 42; and Caplow, *Michelozzo*, 1:186, n. 74, where the author says that Pope-Hennessy considered the middle Virtue as Donatello's. I would like to thank Prof. J. Raspi-Serra of the University of Salerno for helping me to make arrangements to see the Naples tomb.

16. Ph.D. candidate Maria Pernis in a seminar at Columbia University held in 1985, the result which will be published soon in *Source* magazine, made a distinction in Donatello's experience with Classical sculpture, from a dependence upon Roman examples up through the twenties and early thirties, and a direct acquaintance with original (in distinction to Roman copies) Greek marbles, then being imported into Italy for private collectors, including Cosimo. Thus the head of the central Caryatid could be compared with the head of Mary in the *Cavalcanti Annunciation* in Santa Croce, which is Hellenic rather than Roman in flavor.

17. Caplow, *Michelozzo*, 1:204, n. 96.

contrary to widely held opinion. Quite seriously incomplete and with some areas merely blocked out — including the hands — most of the final refinements that Donatello would have applied are entirely absent. Still its powerful, realistic quality need not cause anxiety when considered with the more Classical elements discussed above. We are quite aware today that artists are not restricted to a single mode at any given time, but might easily move back and forth, when the subject required it. They were never as rigid about apparent consistency of style as critics who study their art tend to be.

One final strongly Donatellian component of the Brancacci Tomb is the *tondo* relief of God the Father, with all its coloristic detailing [Fig. 5]. If the tondo as a form had trecento precedents, it became in the quattrocento an important motif for tomb sculpture: one need only think of the Bruni and Marsuppini Tombs in Santa Croce or the Cardinal of Portugal Tomb in San Miniato. In Naples it is set forth somewhat tentatively but prominently, and the low relief technique is at least in the 1420s almost exclusively a Donatellian device.

When significant portions of the Tomb are restored to Donatello, his obligation to the patron, Cosimo de 'Medici, is more easily understood. The deference that Donatello would have shown, if my analysis is correct, toward Cosimo, is confirmed by the the growing intimacy that evolved between the two men over the next decades. To consider the situation in any other way would be to close one's eyes to the remarkable, innovative features of the Tomb and the circumstances of its creation.[18]

■

18. The portions of the Brancacci Tomb ascribed to Donatello herein correspond in large part to Martinelli's observations (see n. 13 above) and are not entirely at odds with Lightbown's conclusions.

APPENDIX I

ASF, MAP, 153, II.

[fol. 13, right side]

+MCCCCXX

Messer Rinaldo cardinale de' Branchacci de'avere adì xxvi di novembre fiorini otomila d'oro, per fiorini VII mila DC di camera cie l'asegniorono per creditore i Medici di Corte di Firenze; posto debi dare in Corte fiorini 7600 F. VIII mila
+Ane scritta in nome di detti Medici, in dì xi di magio 1419

[fol. 26, left]

MCCCCXXII

[1] Messer Rinaldo Cardinale de' Branchacci de'dare adì xi di marzo 1425 [= 1426, modern] f[orini] mille secento novanta, posto Cosimo e Lorenzo de' Medici et C[ompagn]i di Firenze debino avere in questo, c. 39, sono per f[orini] MDC di Camera, per noi pagliorino al detto cardinale e scrìssoli in su la scritta e i nostri di Firenze li fecion loro buoni F. MDCLXXXX

[2] E dì xxiiii di marzo 1428 [= 1429, modern] f[orini] ottocento cinquanta di camera, per lui faciemo buoni a'nostri di Firenze in f. VIIII C [above the line] che chosì li ragionamo, posto li debino avere in questo a c. 47, sono per f[orini] MCCCL di Camera si à pagare per la sua sipultura, chome ordinorono Cosimo de' Medici et C[ompagn]i a Donato di Niccolò e Michelozo di Bartolomeo intagliatori, che ànno tolto a fare detta sipoltura F. DCCCC

[3] E dì xx di magio 1430, fiorini cinquemila quatrocento sesantasette s[oldi] xxvii d[enari] iiii a ffiorino, per f[iorini] V M [above the line] CL di camera, posto i nostri di Roma debino avere in questo, c.48, i quali pagarono per nostro nome per lo detto chardinale, come ordinò per suo testamento e prèsone la chiareza si richiede e riebono la scritta n'avea di nostro.
F. V M [above] CCCCLXVII s. XXVII d. iiii

[4] E dì v d'ottobre 1431 fiorini quatordici s. II d. iii a ffiorino, posto a Cosimo e Lorenzo e Ilarione, a c. 27, per tanto è d'avanzo a questa regione F. XIII s. II d. iii

+Somma fiorini 8072 s.-d. 7 a ffiorino

[fol. 27, left]

+MCCCCXXII

Cosimo e Lorenzo de' Medici e Ilarione de' Bardi per ragion d'avanzi deb dare...

E dì detto [24 March 1426 = 1427, modern] fiorini seicento trentaquatro, posto messer Rinaldo Cardinale de' Branchacci debi avere in questo a c. 40, sono che per lui s'erano pagati qui quando ci fu la Corte, per pigione dela chasa ove stette e per panni fatti tignere a Vinegia in charmisi e per altre spese fatte per lui, che non li à voluti pagare, e però li mettiano e a questo modo si viene a pagare lui istesso la provisioni di danari di suo abiano F. DCXXXIIII

[fol. 47, right]

Chosimo e Lorenzo de' Medici e compagni di Firenze deono avere...

E dì detto [24 marzo 1428 = 1429, modern] fiorini novecento d'oro per fiorini DCCCCL di camera faciemo loro buoni per messer Rinaldo cardinale de' Branchacci, posto li debi dare in questo a c. 26 sono per la sua sipoltura fa fare Cosimo nostro Fior. DCCCC

APPENDIX II

Below I present the details of a case brought against the Operai of Santa Maria at Prato who were in charge of the building of the outdoor pulpit that Donatello and Michelozzo had contracted to make in 1428, that is, during the period when they were finishing up work on the Brancacci monument. Work dragged on for about a decade in Prato. In this case the Florentine carver (*scarpellatore*) Papi di Piero, who has been identified as a partner of Pagno di Lapo by Lightbown (*Donatello and Michelozzo*, 2:244), sued the Pratesi for work he conducted over 180 days spent on the *pergamo* during 1434 and 1435 without being paid, for a total of 18 gold florins.

Pagno di Lapo, through the *portato* prepared in his name by his father for the Catasto of 1427, which was apparently submitted a year late, was owed 18 months salary by the firm of Donatello and Michelozzo. The two masters were then in Pisa, according to the document. (See C. v. Fabriczy, "Pagno di Lapo Portigiani," *Jahrbuch der Königlich Preussischen. Kunstsammlungen* 24 (1903), beiheft, 128, and cited in Janson, *Donatello*, 90, n. 6.

During this period there was a certain laxity in paying the artisans. At the same time it is also known that Pagno was active in Siena at least part of the period when he claimed to be owed money by Donatello and Michelozzo. (See J.T. Paoletti, "The Siena Baptistry Font: A Study of an Early Renaissance Collaborative Program, 1416–1434," Ph.D. dissertation, Yale University, 1967, docs. 172 and 206.) As far as we know the dispute with Pagno was settled without official court action.

[ASF, Mercanzia, no. 1316, fol. 484v]

Adì 8 di lug[i]o [1435]

Dinanzi a voi, messer ufficiale e Corte, expone et dice:

Papi di Piero, scarpellatore del popolo di L. Lorenzo di Firenze, che gl'Operai della capella della pieva [sic] di Prato, che sono diputati sopra fare il pe[r]gamo di fuori di decta pieve, et ciascuno di loro in tucto, furono et sono veri debitori del decto Papi in fiorini diciotto d'oro, per resto de prezo et mercede et salario d'opere cento ottanta, alloro data [sic] in Prato nel decto pergamo, a soldi 16 denari 6 il dì, per lo decto Papi, parte dell'anno 1434 et parte dell'anno presente 1435, come chiaro apparisce et appare debba al libro grande di decti Operai, segnato...[blank], carta...[blank]. Et più volte richiesto che pag[h]inno, ànno cessato pagare, contra al dovere. Et però domanda che piaccia a voi, messer ufficiale e Corte, fare comandare a decti Operai et a ciascuno di loro che infra dieci dì promiximi futuri produchino in iuditio il decto loro libro dinanzi al decto uffficiale et sua Corte, et dieno et paghino al decto Papi la decta quantità di fiorini diciotto d'oro, altrimenti, passato il decto termine gli abbiate per confessi et convincti, et contra a lloro et ciascuno di loro in tuto procediate et facciate secondo la forma degli statutti et ordini di decta Corte, constringendoli al pagamento per ogni rimedio opportuno, et condampnandogli etiandio per sententia ognuno di loro in tucto, se le predecte cose neciessaseno fare. Et adomanda le spese facte et che farà, e ragione e giustitia.

Ad petitione di decto Papi, Lastraio messo di decta Corte rapportò al decto ufficiale et a me notaio infrascritto se' a dì sei di questo mese, di licentia di decto ufficiale, avere richisesti il [sic] decti Operai per ogi a vedere la decta petitione e domanda, torne copia e oporre, alias etc. Ancora rapportò decto messo avergli comandato che infra dieci dì proximi futuri produchino in giudicio il decto loro libro, et dieno et paghino al decto Papi la decta quantità di fiorini diciotto e le spese, alias, etc., in persona a Iacopo...[blank], concedo e a Luccio [?] in persona, e gli altri Operai alle case et alle loro persona di decti Operai.

■

Figure 1. Tomb of Cardinal Rinaldo Brancacci. Naples, Sant'Angelo a Nilo.

Figure 2. Detail of Figure 1. Central Virtue Caryatid. Here attributed to Donatello.

Figure 3. Detail of Figure 1. Central Virtue-Caryatid. Here attributed to Donatello.

Figure 4. Detail of Figure 1. Left Pinnacle Putto. Here attributed to Donatello.

Figure 5. Detail of Figure 1. Tondo of God the Father, detail. Here attributed to Donatello.

XI
Ekphrasis and the Renaissance of Painting: Observations on Alberti's Third Book

DAVID ROSAND

"VEDERAI TRE LIBRI," writes Leon Battista Alberti to Filippo Brunelleschi in the prologue to *Della pittura,* the Italian version of his *De pictura.* "El primo, tutto matematico, dalle radici entro dalla natura fa sorgere questa leggiadra e nobilissima arte. El secondo libro pone l'arte in mano allo artefice, distinguendo sue parti e tutto dimostrando. El terzo instituisce l'artefice quale e come possa e debba acquistare perfetta arte e notizia di tutta la pittura."[1] In thus describing his remarkable and quite unprecedented humanist tract on a mechanical art, Alberti was introducing the great Florentine architect to the Classical form of his book. *De pictura* is structured on the model of the ancient rhetorical treatise, and it is precisely the appropriation of that model for a discussion of the art of painting that serves to ennoble the subject itself.[2]

1. The standard edition of *De pictura* is by Cecil Grayson (Rome-Bari, 1975) — reprinted from Leon Battista Alberti, *Opere volgari,* vol. 3 (Bari, 1973) — which includes facing pages of both the Latin text and Alberti's Italian translation. Citations refer to the paragraph numbers of Grayson's edition; English translations are from Leon Battista Alberti, *On Painting and On Sculpture,* ed. and trans. Grayson (London, 1972), which offers the Latin text as well. Although it is generally assumed that Alberti conceived his treatise in Latin and subsequently translated it into Italian for his artist friends, an argument for the priority of the Italian version has been made by Maria Picchio Simonelli, "On Alberti's Treatises of Art and their Chronological Relationship," *Yearbook of Italian Studies* (1971): 75–102.
2. See Creighton E. Gilbert, "Antique Frameworks for Renaissance Art Theory: Alberti and Pino," *Marsyas* 3 (1943–45): 87–106; and, stressing a different model, D.R. Edward Wright, "Alberti's *De Pictura:* Its Literary Structure and Purpose," *Journal of the Warburg and Courtauld Institutes* 47 (1984): 52–71. See also Michael Baxandall, *Giotto and the Orators: Humanist Observers of Painting in Italy and the Discovery of Pictorial Composition 1350–1450* (Oxford, 1971), 121–39; and John R. Spencer, "*Ut Rhetorica Pictura:* A Study in Quattrocento Theory of Painting," *Journal of the Warburg and Courtauld Institutes* 20 (1957): 26–44.

Alberti develops his theme in three parts: the rudiments (*elementa*), painting (*ars*), and the painter (*artifex*) — transposing the Classical divisions of *poesis, poema,* and *poeta* to the art of painting.[3] Book One, "tutto matematico," offers the earliest account of the new system of perspective construction, grounding pictorial representation in geometry and thereby attesting to the intellectual status of the art of painting as a rational activity of the mind. Book Two moves from the mathematical to the figural, from the perspective schema that is the foundation of spatial illusion in a picture, through the rendering of bodies, to the elements that constitute its affective fiction, its *historia*. Here Alberti discusses the technical procedures of drawing, composition, and coloring, the three basic elements of the art ("Picturam igitur circumscriptio, compositio et luminum receptio perficiunt" [¶31]), as well as the antiquity and nobility of the art. He addresses the rhetorical function of painting, its goal of moving the soul of the beholder, and this theme naturally allows him to pillage the storehouses of ancient anecdotes and *topoi* — especially Pliny's *Natural History.*

After the mathematical challenge of Book One, with its apparently demanding geometry, and the rhetorical riches of Book Two, with its acute analyses of pictorial technique and affect, the final book of *De pictura* may seem a somewhat less impressive achievement, almost an afterthought. "Several things, which I do not think should be omitted from these books, still remain to complete the instruction of the painter, so that he may attain all the praiseworthy objects of which we have spoken. Let me now explain them very briefly" (¶51). So opens Book Three, which is indeed quite brief. And, as if in response to its modest length, it has received relatively little critical attention.[4] The first two books, respectively Euclidean and Ciceronian in tone, have inspired an impressive tradition of scholarship and commentary.[5] Together, they

3. "Rudimenta," "Qui pictura inscribitur," and "Pictor incipit," read the rubrics in the first printed edition: *De pictura praestantissima arte et nunquam satis laudata libri tres absolutissimi, Leonis Baptistae de Albertis...* (Basle, 1540).

4. Baxandall, *Giotto and the Orators,* 125: "Book III, the least substantial, is still the first extended discussion of how painters stand to other artists, particularly writers."

5. See the bibliographies in Grayson's editions: *De pictura,* xxviii–xxxi; *On Painting and On Sculpture,* 27–29; to which should be added: Samuel Y. Edgerton, Jr., *The Renaissance Rediscovery of Linear Perspective* (New York, 1975); and Heiner Mühlmann, *Äesthetische Theorie der Renaissance: Leon Battista Alberti* (Bonn, 1981). Consideration of Alberti's mathematics and optics has more recently extended to his treatment of light and color: see Edgerton, "Alberti's Colour Theory: A Medieval Bottle without Renaissance Wine," *Journal of the Warburg and Courtauld Institutes* 32 (1969): 109–34; James

articulate two discourses considered axiomatic for any definition of definition of Renaissance painting: mathematically-based perspective construction and the classically controlled imitation of natural form. To this major cultural agenda Book Three may appear to add little: the injunction that the painter be morally upright and learned ("in primis esse virum et bonum et doctum bonarum artium" [¶52]) — a humanist inflection, really, of a traditional exhortation, sounded, for example, by Cennino Cennini[6] — and that he frequent the company of poets, philosophers, and orators; there are specific suggestions for subject matter, some further comments on practice, additional anecdotes about ancient painters. Only the concluding paragraph has appealed to the modern historical imagination with any particular force: Alberti's final plea that subsequent painters reward him for his labors by including his portrait in their *historiae* and his most Renaissance self-acknowledgment: "I consider it a great satisfaction to have taken the palm in this subject, as I was the first to write about this most subtle art" (¶62).

Book Three is devoted to the painter, whose goal should be to achieve praise and fame rather than riches through his work (¶52). Having thus distinguished the painter's art as liberal rather than mechanical, Alberti proceeds to urge his association with the *litterati*.[7] Although the painter must first of all have a solid knowledge of geometry, the mathematical foundation of his art, he should be learned as well in all the liberal arts, which will offer him its substance.[8] Literary knowledge, acquired presumably from his poet

---

S. Ackerman, "Alberti's Light," in *Studies in Late Medieval and Renaissance Painting in Honor of Millard Meiss*, ed. Irving Lavin and John Plummer (New York, 1977), 1–27; and idem, "On Early Renaissance Color Theory and Practice," *Memoirs of the American Academy in Rome* 35 (1980): 11–44 (= *Studies in Italian Art History* 1).

6. Cennino d'Andrea Cennini, *Il libro dell'arte,* ed. Daniel V. Thompson, Jr. (New Haven, 1932), ch. xxviiii; cf. also chs. i, ii.

7. On this theme, see Rudolf Wittkower, *The Artist and the Liberal Arts* (London, 1952); and, for the larger context of the issue, Paul Oskar Kristeller, "The Modern System of the Arts" (1951–52), reprinted in his *Renaissance Thought and the Arts: Collected Essays* (Princeton, 1980), 163–227. But cf. Baxandall's reaction to what he apparently considers overgeneralized assumptions in modern historical writing: "Whether or not painting was a liberal art is mercifully not an important theme of the early humanists, but neither were other kinds of conjecture about its status" (*Giotto and the Orators,* 122). Three pages later, however, he himself quotes Pier Paolo Vergerio (1404): "Designativa vero nunc in usu non est pro liberali..." (ibid., 125).

8. Here Alberti's model very likely was Vitruvius, *De architectura,* I.1, the opening discourse on the education of the architect — e.g.: "Et ut litteratus sit, peritus graphidos, eruditus geometria, historias complures noverit, philosophos diligenter audierit, musicam scierit, medicinae non sit ignarus, responsa iurisconsultorum noverit, astrologiam caelique rationes cognitas habeat" (1.1.3).

and orator friends, will assist the painter in composing his *historiae*. It is indeed the *historia*, the true test of the painter's powers of invention, to which Book Three returns as the most important part of the painter's work.

Continuing his course for the training of the painter, and addressing themes that will remain constants in the subsequent literature on art at least through the eighteenth century, Alberti shifts easily between theory and practice, general aesthetics and studio technique. He compares the progress of the novice artist to that of a youth learning to write: beginning with the basic elements of the alphabet, the pupil moves on to form syllables and then words; just so should the artist progress from drawing simple outlines to more complex structures.[9] Nature should be the guide and its variety studied and emulated, but ultimately its model must be improved by nobler aesthetic considerations; it must be perfected by an added beauty (¶55). Beauty, however, is not readily grasped; it is not to be found in any single model — hence the story of Zeuxis and the five maidens of Croton (¶56). Alberti appeals to an idea of beauty (*pulchritudinis idea*) in the mind of the artist, an idea the artist strives to realize in his work.[10]

Nonetheless, beauty can be rendered only when the artist has learned from Nature. It is Nature, depicted as a reflection of ourselves, that lends special affective power to painting:

9. "I would have those who begin to learn the art of painting do what I see practised by teachers of writing. They first teach all the signs of the alphabet separately, and then how to put syllables together, and then whole words. Our students should follow this method with painting. First they should learn the outlines of surfaces, then the way in which surfaces are joined together, and after that the forms of all the members individually..." (¶55). Vergerio's comment, quoted above (note 7), reads more fully: "Nowadays drawing does not in practice pass as a liberal study except so far as it relates to the writing of characters — writing being the same thing as portraying and drawing — for it has otherwise remained in practice the province of the painters" (trans. Baxandall, *Giotto and the Orators,* 125). Alberti's pedagogic analogy is applied more directly to the handling of the pen by Giovan Battista Armenini, *De' veri precetti della pittura* (Ravenna, 1587), I.vii, in Paola Barocchi, ed., *Scritti d'arte del cinquecento* (Milan-Naples, 1973), 1:2006.

10. For the larger context of Alberti's notion, see Erwin Panofsky, *Idea: A Concept in Art Theory* (1924; 2nd ed., 1960), trans. Joseph J.S. Peake (Columbia, SC, 1968), esp. 57–58.

> ...the painter who has accustomed himself to taking everything from Nature will so train his hand that anything he attempts will echo Nature. We can see how desirable this is in painting when the figure of some well-known person is present in a *historia*, for although others executed with greater skill may be conspicuous in the picture, the face that is known draws the eyes of all spectators, so great is the power and attraction of something taken from Nature (¶56).

Masaccio had already demonstrated this power to most moving effect in his *Trinity* in S. Maria Novella. Alberti, extracting principle from practice, anticipates and invites what will become common in Renaissance painting, the inclusion of portraits (usually of donors) in religious and historical painting. It is, of course, the principle that he will invoke at the conclusion of *De pictura* on behalf of his own claim to immortality — "Through painting," as he declared at the beginning of Book Two, "the faces of the dead go on living for a very long time" (¶25).[11]

Still more striking is Alberti's anticipation of other practices that would become common in Renaissance studios, especially concerning the role of drawing in training and in the preparation of compositions. The discussion touches on the copying of works by other masters — traditional procedure, fully addressed by Cennini[12] — and drawing after sculpture, which in turn verges on, without actually achieving, a pioneering formulation of the *paragone* between painting and sculpture (¶58).[13] Of greater interest, however, is Alberti's commentary and advice on the technical procedures of drawing and painting, which take him yet closer to the problems of studio practice. Draw large, he urges, as large as life, so as not

11. Alberti's defense of the value of painting begins: "Painting possesses a truly divine power in that not only does it make the absent present (as they say of friendship), but it also represents the dead to the living many centuries later, so that they are recognized by spectators with pleasure and deep admiration for the artist." Cf. Cicero, *De amicitia,* 7.23. For further comment on this theme in practice, see David Rosand, "The Portrait, the Courtier, and Death," in *Castiglione: The Ideal and the Real in Renaissance Culture,* ed. Robert W. Hanning and David Rosand (New Haven-London, 1983), 91–129.
12. Cennini, *Il libro dell'arte,* ch. xxvii.
13. On the *paragone* of the arts, which emerges as a full topic about 1500, see Irma A. Richter, *Paragone: A Comparison of the Arts by Leonardo da Vinci* (London, 1949). For the issue in the art of the earlier Renaissance, see John White, "Paragone: Aspects of the Relationship between Sculpture and Painting," in *Art, Science, and History in the Renaissance,* ed. Charles S. Singleton (Baltimore, 1967), 43–109.

to hide your errors (¶57). As he moves his argument along, Alberti raises issues that often will become pressing only a generation or more later: for example, the dialectic tension between diligence and speed of execution — his own Italian, naturally, comes closer to the tone of the later shop vernacular: "bisogni avere una diligenza congiunta con prestezza" (¶59). These should be combined, he explains, "in such a way that the painter will never apply his brush or style to his work before he has clearly decided in his own mind what he is going to do and how he will do it. It is safer to remove errors with the mind than to erase them from one's work."

From the beginning of *De pictura* (¶1) Alberti had insisted that he was writing as a painter, and the conviction of his text depends precisely upon his ability to move so comfortably between theory and practice, between the literary conventions of his humanist training, which enabled the project, and the technical problems of the studio, which occasioned it. His discourse on *diligentia* and *celeritas*, supported of course by appropriate Plinian anecdote, continues his brief for the intellectual basis of the art of painting. "Talent roused and stimulated by practice turns easily and readily to work, and the hand swiftly follows when guided by sure judgement.... [The artist] should never put his hand to work without the guidance of well-informed judgement" (¶59) — "la man che ubbidisce all'intelletto," as Michelangelo would write a century later.[14] Alberti does not linger long on that elevation, nor does he give any thought to its Platonic implications. Rather, he returns immediately to practical matters, to the "variety and abundance, without which no *historia* merits praise" (¶60) and to the preparation of the composition.

> When we are about to paint a *historia*, we will always ponder at some length on the order and the means by which the composition might best be done. We will work out the whole *historia* and each of its parts by making sketch models on paper, and take advice on it with all our friends (¶61).

Alberti's advice on working out the composition in preparatory drawings — antedating the earliest extant compositional sketches of the Italian quattrocento — probably coincides with a growing contemporary practice, but his rationale would seem to add a larger dimension to that practice. Indeed, we are tempted to ascribe to him

14. Michelangiolo Buonarotti, *Rime*, ed. Enzo Noè Girardi (Bari, 1967), no. 151.

responsibility for the invention of *disegno* in its Renaissance — that is, its modern — sense.[15]

His concern with rationalizing the conceptual procedures of pictorial invention forms part of the same program as his rationalization of pictorial illusion through a rigorously mathematical perspective construction.

> We will endeavour to have everything so well worked out beforehand that there will be nothing in the picture whose exact collocation we do not know perfectly. In order that we may know this with greater certainty, it will help to divide our models into parallels, so that everything can then be transferred, as it were, from our private papers and put in its correct position in the work for public exhibition (¶61).

Alberti here offers us the earliest statement on the principle and function of the cartoon, the full-size preparatory drawing that will replace the traditional *sinopia* in the preparation of frescoes.[16]

His program throughout *De pictura* involves the rationalization of pictorial practice in all its dimensions, from conception to execution. This is what renders painting an intellectual pursuit, the business of the mind. By defining the creative process, that is, the inventive part of painting, as essentially graphic, displacing it from the wall to separate sheets of paper, Alberti was being true to his own sense of the art.

Interestingly, his tripartite division of painting into *circumscriptio, compositio,* and *luminum receptio,* although echoing in form the antique literary model *(inventio, dispositio, elocutio),* responded in substance to the more technical exigencies of pictorial practice. According to the pedagogical intentions of *De pictura,* practice meant first of all the delineation, construction, and combination of form, the

15. This thesis will be argued in my forthcoming book, *On Drawing: Critical and Historical Studies.* A recent survey of the material is Francis Ames-Lewis, *Drawing in Early Renaissance Italy* (New Haven-London, 1981).
16. Squaring for enlargement and transfer had already been put into practice by Masaccio in the *Trinity* in S. Maria Novella. On the preparation of mural painting, see Ugo Procacci, *Sinopie e affreschi* (Florence, 1960); Millard Meiss, *The Great Age of Fresco: Discoveries, Recoveries, and Survivals* (New York, 1970); Eve Borsook, *The Mural Painters of Tuscany from Cimabue to Andrea del Sarto,* 2d ed. (Oxford, 1980). For the continuing debate regarding the role of drawings, see Procacci, "Disegni per esercitazione degli allievi e disegni preparatori per le opere d'arte nella testimonianza del Cennini," in *Studies...in Honor of Millard Meiss,* 352–67, with references to the preceding bibliography; also, in the same volume: Umberto Baldini, "Dalla sinopia al cartone," 43–47.

building up from planes to members to bodies (¶35), from outline to composition to modelling.[17] The concerns of Alberti the painter more than balance the habits of Alberti the humanist. The fundamentally graphic nature of his subject, after all, had led him to the more elementary and practical comparison with writing. That comparison, in turn, was prefaced by a declaration of principle: "The fundamental principle will be that all the steps of learning should be sought from Nature" (¶55). A similar appeal to Nature had been made in support of his tripartite definition of painting: "We divide painting into three parts, and this division we learn from Nature herself " (¶30). Whether or not Alberti felt the need to apologize for his deviation from the ancient model, the invocation of Nature as mistress sounds a theme that will play an especially significant part in the pictorial (as well as poetic) aesthetics of the Renaissance. Considered against the inherited literary conventions of his treatise, Alberti's appeal to Nature becomes in effect an appeal to experience and to practice — in this, as in so much, anticipating that "disciple of experience," Leonardo da Vinci.[18]

The Classical rhetorical division of *inventio, dispositio,* and *elocutio* ordered the development of the finished work from the selection of material through the general organization of the composition to the final execution or presentation. It would later be more exactly reflected in the formulation and vocabulary of cinquecento art theory: *invenzione, disegno, colorito.*[19] And yet,

17. Regarding rhetoric vs. practice in Alberti, Mark W. Roskill has remarked: "The first Renaissance theorist to formulate such a scheme for painting on the antique model had been Alberti. But the three components which he had singled out...accorded with the basic mechanics of pictorial execution; literary theory was not yet involved." See *Dolce's "Aretino" and Venetian Art Theory of the Cinquecento* (New York, 1968), 267. Roskill's perception seems to us basically accurate, although his conclusion is contradicted by the very tone and ambition of *De pictura.* Further comment on this theme: Emma Barilla, "The 'Sister Arts' in Alberti's 'Della Pittura'," *British Journal of Aesthetics* 19 (1979): 251–62.

18. *The Notebooks of Leonardo da Vinci,* ed. Edward MacCurdy (New York, 1939), 989 (C.A. 191r.a), discussed by Carlo Pedretti, *Leonardo: A Study in Chronology and Style* (Berkeley-Los Angeles, 1973), 15.

19. Cf. Paolo Pino, *Dialogo di pittura* (1548), ed. Paola Barocchi in *Trattati d'arte del cinquecento,* (Bari, 1960), 1:113; followed by Lodovico Dolce, *Dialogo della pittura, intitolato l'Aretino* (1557), ed. Barocchi, 164. For further discussion: Gilbert, "Antique Frameworks," esp. 94–106; Rensselaer W. Lee, *Ut Pictura Poesis: The Humanistic Theory of Painting* (1940; New York, 1967), esp. 70–71; and the commentaries in Barocchi, *Trattati d'arte*; and Roskill, *Dolce's "Aretino."*

even though Alberti evidently felt obliged to modify the formula to accommodate actual pictorial practice — and especially, we may assume, the absolutely novel phenomenon of mathematical perspective — the literary values remain central to his discourse. Along with the geometry of perspective, they are precisely what set the higher goals and standards for painting as a noble art. *Inventio,* although absent from his tripartite definition of painting, runs through *De pictura,* especially Book Three, as the essential virtue, the quality that distinguishes the painter as a creative mind.

"The great work of the painter is the *historia,*" Alberti declares on several occasions (¶33, 35), and "the great virtue of this consists primarily in its invention. Indeed, invention is such that even by itself and without painting it can give pleasure" (¶53). It is, understandably, in his discussion of the *historia* that Alberti has most direct recourse to the rhetorical terms of antiquity. *Inventio* here exists effectively on the level of idea, beyond the contingency of pictorial execution. It involves the "what" rather than the "how" of representation, the latter being the concern of Alberti's basic, essentially technical, definition of painting. Invention is the imaginative act of conceiving an *historia,* of finding appropriate material to be represented, and it is the reason that painters should be literate and the companions of *litterati.*

Historia and *inventio* belong to the realm of literary discourse. By introducing them into his discussion of painting Alberti deliberately expands the dimensions of that art, enlarging its cultural reach and the very nature of its relationship to an audience. Significantly, this is the aspect of his commentary that is totally without precedent in contemporary practice. Herein, then, lies the profound originality of *De pictura,* its proleptic resonance.

To demonstrate the inherent interest of a pictorial invention without painting, Alberti offers the first of two examples:

> The description that Lucian gives of Calumny painted by Apelles excites our admiration when we read it. I do not think it is inappropriate to tell it here, so that painters may be advised of the need to take care in creating inventions of this kind (¶53).

Alberti proceeds to describe the long lost painting by the most celebrated painter of antiquity, a painting he, of course, knew only through its description in Lucian's *De calumnia* — more precisely, through Guarino's translation of Lucian's *ekphrasis*.[20] "If this *historia* seizes the imagination when described in words," he concludes, "how much beauty and pleasure do you think it presented in the actual painting of that excellent artist?" With a casualness bordering on the disingenuous, Alberti has sounded perhaps the most resonant note of his entire treatise, a programmatic call to painting that — as much as the commensurability of perspective, Classical figural proportion, or the orders in architecture — must be counted central to any concept of the Renaissance in art.

Following his *ekphrasis* of the "Calumny of Apelles," Alberti presents another, of the Three Graces:

> What shall we say too about those three young sisters, whom Hesiod called Egle, Euphronesis and Thalia? The ancients represented them dressed in loose transparent robes, with smiling faces and hands intertwined; they thereby wished to signify liberality, for one of the sisters gives, another receives and the third returns the favour, all of which degrees should be present in every act of liberality. You can appreciate how inventions of this kind bring great repute to the artist. I therefore advise the studious painter to make himself familiar with poets and orators and other men of letters, for he will not only obtain excellent ornaments from such learned minds, but he will also be assisted in those very inventions which in painting may gain him the greatest praise. The eminent painter Phidias used to say that he had learned from Homer how best to represent the majesty of Jupiter. I believe that we too may be richer and better painters from reading our poets, provided we are more attentive to learning than to financial gain (¶54).

20. "In the painting there was a man with enormous ears sticking out, attended on each side by two women, Ignorance and Suspicion; from one side Calumny was approaching in the form of an attractive woman, but whose face seemed too well versed in cunning, and she was holding in her left hand a lighted torch, while with her right she was dragging by the hair a youth with his arms outstretched towards heaven. Leading her was another man, pale, ugly and fierce to look upon, whom you would rightly compare to those exhausted by long service in the field. They identified him correctly as Envy. There are two other women attendant on Calumny and busy arranging their mistress's dress; they are Treachery and Deceit. Behind them comes Repentance clad in mourning and rending her hair, and in her train chaste and modest Truth." On the entire subject, see David Cast, *The Calumny of Apelles: A Study in the Humanist Tradition* (New Haven-London, 1981), with earlier bibliography.

Ekphrasis, the ancient rhetorical exercise in description, had been an important part of the early humanists' response to painting. Guarino and his school, in particular, had delighted in the detailed variety of paintings — such as those of Pisanello — which provided the occasion for literary performance, the display of rhetorical prowess.[21] The copiousness of a painting, more concentratedly abundant than nature's, proved an irresistible challenge to the writer's descriptive skills. Michael Baxandall has very elegantly shown just how Alberti differed from his humanist colleagues, how he developed the conventional values of rhetoric into a more systematic response to and program for painting, looking beyond the descriptive detail and creating, in effect, the very notion of pictorial composition.[22] But Alberti hardly abandoned the ekphrastic mode. Rather, he inflected it so that, no longer merely responsive, it acquired new relevance and a new function: pro- as well as retrospective, it became both the historical record of past achievement in painting and the source of future achievement. Instead of indulging in a verbal tapestry to match the visual riches of the International Gothic style, he recalled the interpretive ambitions of the ekphrastic traditions of antiquity, intending to challenge intellect as well as sense. Beyond their descriptive function, Alberti's *ekphraseis* were prescriptive as well; he offered them as bridges across time, across the Middle Ages, as *exempla* linking the glory of Classical antiquity and a pictorial Renaissance that he was effectively inventing. Only through such literary accounts were the monuments of ancient painting preserved; only their *inventioni* survived the ravages of time.

The verbal representations of the "Calumny of Apelles" and of the Three Graces are explicitly offered as models, inventions for *historiae*. From these descriptions the contemporary painter, Alberti's intended reader, is invited to reconstruct images from the Classical past. It is indeed a grand invitation: to join Alberti in the project of creating nothing less than the full Renaissance of painting. It was an invitation that could not be immediately accepted, however.

The first two major parts of Alberti's curriculum, mathematically-based perspective and rhetorical naturalism, existed in easy harmony with the pictorial culture that Alberti discovered in Florence on his

21. On the ekphrastic mode among the humanists, see Baxandall, *Giotto and the Orators,* 78–96.
22. Ibid., 121–39 (ch. III: "Alberti and the Humanists: Composition").

return from exile. Brunelleschi, Ghiberti, Donatello, and Masaccio had brought perspective construction under rigorous control in practice. In figural design as well as in architecture, they had recovered the commensurability of Classical proportion, and they had civilized the Christian pathos of late Gothic imagery with the *gravitas* of Classical form. With regard to the "how" of representation, then, Alberti might be said to have articulated the achievements of contemporary practice. His recommendations regarding the "what" of representation, on the other hand, however logical they may appear to us in historical hindsight, were fundamentally novel, advanced in every sense.

The absence of any reference to Christian subject matter in *De pictura* has always struck modern readers of Alberti — the illustration of Giotto's *Navicella* provides only an apparent exception, for that picture is cited for its dramatic expression of the *affetti* (¶42). Whatever the enthusiasm of his dedicatory letter to Brunelleschi, Alberti remains frustratingly silent on the actual state of painting in Florence about 1435 — although his criticism of the application of gold leaf (¶49), a violation in every way of his concept of the picture plane as an intersection of the pyramid of vision, suggests something of the strength of his attitude. *De pictura* — and especially, we may imagine, *Della pittura* — was clearly intended to do more than celebrate the status quo: "di nuovo fabrichiamo un'arte di pittura" (¶26). Alberti had a program in mind for the further reform of painting, one that would build upon the foundations provided by the heroic generation of the first quarter of the quattrocento. With critical foresight and real imagination he understood in their fullest significance the implications of what Donatello and Masaccio had achieved, and he was determined to see that potential realized. It is that determination that motivates Book Three.

Book Two, in which the individual elements that constitute the *historia* are discussed, offers a repertoire of expressive motifs inspired by antiquity; each becomes the occasion for and illustration of a larger point of principle — as the description of the dead Meleager being carried away (an *historia* they praise in Rome) initiates a discourse on the decorum of representation (¶37). And even without explicit reference, we intuit the ancient models behind his further comments on the expressive movement of inanimate things, "of hair and manes and branches and leaves and clothing" (¶45). The flow of hair and draperies (which Aby Warburg would characterize as "bewegtes Beiwerk") gives graphic articulation to the movements of the body, sustaining those established gestures of passion (Warburg's "Pathosformeln") that Alberti had certainly observed on ancient Roman sarcophagi and had seen animated to new purpose in the work of Donatello.[23] Within the controlled setting of perspective construction, against the measured grid of the plane and the proportionality of its forms, the *historia*, after all, was expected to move the passions of the beholder. Such volatility was not to be found in the Giottesque gravity of Masaccio, nor would it be in the work of the painter who is so often cited as the ideally Albertian master, Piero della Francesca. Rather, Alberti's call for dramatic movement, for graphic correlatives to human expression, calls up the work of a painter like Castagno or, still more relevant, of artists of the next generation, Antonio Pollaiuolo, Botticelli, and beyond Florence, in north Italy, Mantegna.

From the individual elements and motifs of Book Two, Alberti expands the discussion in Book Three to embrace the full *historia*, offering as models the two *ekphraseis*. His selection is neither casual nor arbitrary. The topics, in fact, are highly professional; each presents a particular challenge and opportunity to the modern painter. The "Calumny of Apelles" is more than an elaborate morality play *all'antica* with a *dramatis personae* of colorful personifications; it illustrates a major theme of Alberti's third book: the honor of the painter. Lucian's own description of Apelles' painting, intended to introduce the embodiment of the vice that was his topic, was prefaced by an account of the event that inspired the original picture. Apelles had been wrongly accused by another artist of having participated in a plot against Ptolemy IV Philopater in Egypt and condemned to death. The painter was vindicated just in time by late testimony. Justice prevailed, but Apelles recorded the bitter experience in his allegorical composition, showing the origins and effectiveness of Calumny in the world — but also the eventual triumph of Truth.[24] Lucian's account added a special kind of biographical

23. Warburg first recognized the significance of these issues in his thesis, "Sandro Botticelli's 'Geburt der Venus' und 'Frühling'" (1893), reprinted in his *Gesammelte Schriften* (Leipzig-Berlin, 1932), 1–58 — discussing Alberti's comments on 10–13. For further discussion, see E.H. Gombrich, *Aby Warburg: An Intellectual Biography* (London, 1970), 177–85, 231–38, 244–51. Recent objections to Warburg's perception ignore Alberti and the deliberately programmatic nature of his observations: cf. Paul Holberton, "Of Antique and Other Figures: Metaphor in Early Renaissance Art," *Word and Image* 1 (1985): 31–58, esp. 32–33.
24. On the iconographic role of the "Calumny of Apelles" in the formation of "naked Truth" personified, see Erwin Panofsky, *Studies in Iconology: Humanistic Themes in the Art of the Renaissance* (1939; New York, 1962), 157–59.

substance to the many anecdotes about Apelles to be gathered from the pages of Pliny. Above all, however, it thrust the painter into the public world as the victim of envy, a situation not likely to go unappreciated in the intensely competitive world — especially the artistic world — of early Renaissance Florence and which would play an increasingly significant role in the adoption of the "Calumny of Apelles" by later artists.[25]

Celebrated as the greatest of the ancient Greek painters, specially privileged by Alexander the Great, Apelles emerged as the model for the new artist of the Renaissance. No higher praise could be accorded a painter than the title of the "new Apelles," a tag that would become well worn by the end of the Renaissance tradition, but one that must have carried a very particular charge at the beginning.[26] Alberti's choice of the "Calumny of Apelles" as his first *ekphrasis* can only appear as most knowing and calculated, professionally motivated.

His second *ekphrasis* is not actually based on a particular work. Evoking Hesiod's nomination of the Graces, Alberti offers a description of how they were represented in antiquity, a general reference but with enough detail to satisfy his intention. Here too, the subject carries its ethical meaning, for the choreography of the three sisters represents the action of giving and receiving that signifies liberality.[27] On one level, at least, we can appreciate the appeal (however indirect) of this interpretation to the dedicatee of *De pictura.* Alberti's brief letter to Gianfrancesco Gonzaga of Mantua sounds the familiar note of the humanist in search of patronage. Indeed, it is the very epitome of that situation, concluding with a direct presentation of himself:

25. Especially Federigo Zuccaro, whose case and its late Renaissance context are reviewed in Cast, *The Calumny of Apelles,* 121–58.

26. Evidently the first modern painter to be buried as *alter Apelles* was Fra Angelico (the inscription in S. Maria sopra Minerva is recorded by Giorgio Vasari, *Le vite...*, ed. Gaetano Milanesi, 9 vols., Florence, 1906, 2:522), although Petrarch had already compared Simone Martini to Apelles and Boccaccio had paid a similar compliment to Giotto. For a survey of "Apelles and the Tradition of the Academies," with further references, see Cast, *The Calumny of Apelles,* 159–96. For the larger context of the Apelles *topos*, see André Chastel, *Art et humanisme à Florence au temps de Laurent le Magnifique* (Paris, 1961), esp. 91–105; and Erwin Panofsky, *Renaissance and Renascences in Western Art,* 2d ed. (Stockholm, 1965), 21–35, 182–88 et passim.

27. On "Seneca's Graces," see Edgar Wind, *Pagan Mysteries in the Renaissance*, rev. ed. (Middlesex-Baltimore-Victoria, 1967), 26–35.

> You could know my character and learning, and all my qualities best, if you arranged for me to join you, as I indeed desire. And I shall believe my work has not displeased you, if you decide to enrol me as a devoted member among your servants and to regard me as not one of the least.

In the context of *Della pittura*, that is, of the book's address to the professional painter, the *ekphrasis* of the Three Graces invokes a number of studio *topoi*. Their loose transparent garments ("soluta et perlucida veste ornatas," "con la vesta scinta e ben monda") were clearly intended as a challenge to the painter. Pliny (*Nat. Hist.* 35.58) had celebrated Polygnotos as the first to have painted women with transparent garments (*muliere tralucida veste*), and the depiction of draperies that reveal, in sculpture as well as painting, would become a favorite Renaissance *dimostrazione dell'arte* — especially, of course, in the representation of the female figure. In this too, Alberti's program would eventually be fully realized.

Implicit in Alberti's allusion to the way the Graces were traditionally painted is the juxtaposition of female figures seen alternately from the front and from behind, the kind of juxtaposition that was becoming standard in practice in early quattrocento painting, and not only for the human form.[28] Juxtaposed views allowed the painter a more complete presentation of the figure, the possibility of transcending the limits of his flat plane of operation and picturing his subject in several dimensions. It is the kind of practice that would eventually provide a basic point for the painter's argument in the *paragone* debate.

"Try as we will to illustrate Alberti's *della Pittura* by works of his own time, the images which it conjures up in the mind's eye all come from the painting of the next century."[29] However understandable, Kenneth Clark's observation must seem somewhat off the mark. Whether or not the immediate impact of Alberti's treatise can be seen in the spatial conception and figural organization of Fra Angelico's San Marco altarpiece or could once have been seen in

28. In compositions of the Adoration of the Magi, for example, it was almost traditional to present several views of the horses in the regal train; and Pisanello, in studying the animal, would draw it both head on and from the rear.

29. Kenneth Clark, "Leon Battista Alberti on Painting" (1944), in his *The Art of Humanism* (New York, 1983), 96.

Domenico Veneziano's (lost) frescoes in Sant'Egidio, whether or not *De pictura* inspired new attention to the nude body and Classical statuary, it is clear that its text gave voice to concerns that were central to quattrocento painters and pictorial culture.[30] Alberti learned much from the Florentine artists he so admired, and he learned as much from his study of ancient art. Responding, with Brunelleschi and with Donatello, to the lessons of Roman antiquities, he understood the larger implications of the formal revival of Classical form, and that understanding informed the broad scope of the program he presented in *De pictura*. It enabled him to look beyond the accomplishments of his own generation and anticipate — indeed, to define — the goals of the next.

To limit consideration of the significance of his treatise to matters of perspective or even figural composition is to miss that breadth of vision. Alberti's *ekphraseis* would be realized on a major scale by a later generation, by artists like Botticelli and Mantegna. Botticelli would recreate the *Calumny of Apelles* [Fig. 1] — thereby becoming in fact a new Apelles — and in his *Primavera* [Fig. 2] he brought to exquisite life that image of "those three young sisters, whom Hesiod called Egle, Euphronesis and Thalia."[31] Mantegna, too, would affirm the vitality of these same images.[32] Moreover, in his conscious development of printmaking, he would realize a still more profound aspect of Alberti's program, demonstrating the painter's fecundity in *inventione*. This was the arena in which the imagination of the painter was to be tested, in which he was to display his originality, his *ingenio*; it was where the pictorial intellect worked with idea, transcending the material limits of its medium — so Raphael the inventor, for example, could leave the execution of his

30. On the presumed influence of Alberti's text on painting, cf. the comments of John R. Spencer in the introduction to his translation, *On Painting* (New Haven, 1956), 29–32; Clark, *The Art of Humanism,* 102–4; Richard Krautheimer, "Fra Angelico and — perhaps — Alberti," in *Studies...in Honor of Millard Meiss,* 290–96.
31. For Botticelli's *Calumny of Apelles,* see Cast, *The Calumny of Apelles,* 29–54, with earlier references. Recent cleaning of the *Primavera* has made Botticelli's Graces all the more appealing: see Umberto Baldini, *La Primavera del Botticelli: Storia di un quadro e di un restauro* (Milan, 1984). On the Graces in particular, cf. the remarks of Holberton, "Of Antique and Other Figures," 46–50.
32. The fullest discussion of this relationship is by Michelangelo Muraro, "Mantegna e Alberti," in *Arte, pensiero e cultura a Mantova nel primo Rinascimento,* Atti del VI Convegno internazionale di studi sul Rinascimento (Florence, 1966), 103–32. Baxandall has argued that "it was Mantegna who produced the visual models of Alberti's *compositio*, models in the strict sense of engravings able to carry patterns of Albertian narrative style into the painters' workshops." See *Giotto and the Orators,* 133.

inventions to the engraver Marcantonio, and viewers of those prints could look beyond the ink on paper to encounter the creative mind of the inventor himself.[33]

Ut pictura poesis: in defining the possibilities of painting and the aspirations of the painter, Alberti introduced the ancient simile — conceptually, if not literally — into the new age's consciousness. As important, he had also inflected it to read *ut pictor poeta.* Harnessing the achievement, ambition, and fame of the painter to those of the *litterati,* he established a new standard of valuation and new goals for the art. Taking seriously, on a professional level, what had been essentially rhetorical in his ancient models, Alberti inaugurated the critical dialectic between image and text that would determine attitudes toward art for the next three centuries and more. And it was the notion and practice of *ekphrasis* on which this aspect of his program pivoted.[34]

Nearly a century after Alberti wrote *De pictura,* the topic of its third book would inspire another Renaissance text, of a different order but still very much on painting and the painter:

> Timagora, Parrasio, Polignoto,
> Protogene, Timante, Apollodoro,
> Apelle, più di tutti questi noto,
> E Zeusi, e gli altri ch'a quei tempi foro;
> De'quali la fama (malgrado di Cloto,
> Che spense i corpi, e di poi l'opre loro)
> Sempre starà, finchè si legga e scriva,
> Mercè degli scrittori, al mondo viva:

Representing the poet's brief, the opening stanzas of Canto XXXIII of the *Orlando Furioso* resume the discourse initiated by Alberti, reaffirming his vision of the historical mechanics of the Renaissance of painting.

■

33. On the reading of prints, see David Rosand, "Raphael, Marcantonio, and the Icon of Pathos," *Source: Notes in the History of Art* 3.2 (1984): 34–52. The special role of *inventio* in the development of Renaissance printmaking has been studied by Patricia Emison, "Invention and the Italian Renaissance Print: Mantegna to Parmigianino" (Ph.D. dissertation, Columbia University, 1985).
34. Still a rewarding study of the subject is Lee, *Ut Pictura Poesis.* For the personal inflection of the simile, see David Rosand, "*Ut Pictor Poeta:* Meaning in Titian's *Poesie,*" *New Literary History* 3 (1971–72): 527–46. The fundamental studies of the ancient *ekphrasis* in the Renaissance are those of Richard Förster, which are listed in the bibliography of the most recent contribution to the subject: Michaela J. Marek, *Ekphrasis und Herrscherallegorie: Antike Bildbeschreibungen im Werk Tizians und Leonardos,* Römische Studien der Bibliotheca Hertziana 3 (Worms, 1985), 142.

Fig. 1 .Botticelli, *The Calumny of Apelles.* Gallerie degli Uffizi, Florence. Museum photo.

Fig. 2. Botticelli, *La Primavera*. Gallerie degli Uffizi, Florence. Museum photo.

XII
The Role of Robert Fabyan in Tudor Historiography of the "Wars of the Roses"

J.M.W. BEAN

OUR PRESENT KNOWLEDGE of the role of Robert Fabyan in the evolution of the interpretation of the "Wars of the Roses" that had emerged by the reign of Elizabeth I rests on arguments propounded by A.H. Thomas and I.D. Thornley in their edition of *The Great Chronicle of London,* published in 1938.[1] They argued that both this text and *The New Chronicles of England and France*[2] were written by the London alderman, Robert Fabyan (d. 1513). The attribution of the latter to Fabyan is itself not absolutely certain: it was published anonymously without any attribution to an author by Richard Pynson in 1516, and the ascription to Fabyan was not made until a second edition was published by William Rastell in 1533. It must be said, however, that a number of sixteenth-century authorities, including one — John Stow — who was especially learned in London history, agree on Fabyan's authorship.[3] Moreover, there is independent evidence that he was engaged in studying the history of France, since he borrowed from the Guildhall and retained for a long time "the grete boke of the Croniques Wreton in Frensh."[4]

The theory that Fabyan also compiled *The Great Chronicle* rests in part on paleographical evidence. The two manuscripts of *The New*

1. *The Great Chronicle of London,* ed. A.H. Thomas and I.D. Thornley (London, 1938). The arguments set out in the Introduction are generally accepted. See, e.g., M. McKisack, *Medieval History in the Tudor Age* (Oxford, 1971), 94, n. 1; A. Gransden, *Historical Writing in England, 2: C.1307 to the Sixteenth Century* (London and Ithaca, NY, 1982), 231–32.
2. *The New Chronicles of England and France by Robert Fabyan, Named by Himself the Concordance of Histories,* cited hereafter in the edition of Sir Henry Ellis (London, 1811).
3. *Great Chronicle,* xli.
4. Ibid., xlvi.

Chronicles from which Pynson printed his edition of 1516 are in the same hand as the manuscript of *The Great Chronicle*.[5] In itself this is not totally compelling, since the possibility exists that the manuscripts were copied by the same scrivener for different customers. But three later sixteenth-century authorities all agree that Fabyan was the author of *The Great Chronicle*.[6] The testimony of John Stow is especially important, since the manuscript was already in his possession before 1576 and is annotated by him in the light of a knowledge of London history that was superior to that of any contemporary.

There is one objection to the theory propounded by Thomas and Thornley to which they gave substantial attention. In his account of the troubles of the London alderman Sir Thomas Cooke that began in June 1468 the author makes it clear that he wrote from personal knowledge acquired in his youth: "ffor In the tyme of his ffyrst trowble I was his apprentyze and abowth the age of xvii or xviii yerys and thereabowth."[7] The difficulty is that the Wardens' accounts of the Drapers' Company to which Cooke belonged give no help, since they are incomplete for this period and the only one of Cooke's apprentices they record was not Fabyan. Fabyan himself is recorded as having served his apprenticeship under another merchant. It is possible, however, that he changed masters when Cooke's fortunes collapsed.[8]

The value of *The Great Chronicle* to historians of fifteenth-century England lies mainly in the portion from 1439 onwards since it was the work of a single author who was, at least from the mid-sixties, a contemporary of the events he was narrating. And there can be no doubt that Robert Fabyan who must have been born c. 1450 and died in 1513, if not the same man, was his contemporary. According to Thomas and Thornley, there are no substantial differences between the two narratives.[9] An examination of both, however, reveals that they differ to a great extent in the ways they handled the years in which the government of Henry VI broke down and civil war developed, leading to the accession of the Yorkist Edward IV. Unlike *The Great Chronicle,* from 1450 onwards *The New Chronicles* attributes a major share of responsibility to Henry VI's queen,

5. Ibid., xlvii–lxv.
6. Ibid., xl.
7. Ibid., 205.
8. Ibid., xl–xli.
9. Ibid., lxv–lxvi.

Margaret of Anjou. What follows is a comparison that sets side by side the two chronicles' treatment of the major episodes in the years 1450-59.

The Great Chronicle	*The New Chronicles*
1450–51	
Then the kyng heryng of the dyscumfytuyr of the Staffordys Removyd ffrom Grenewych unto london, and afftyr from london to kyllyngworth [p. 182].	In this whyle, the kynge and the quene herynge of the encreasynge of his rebellys, and also the lords ferynge theyr owne seruantes, remoued from London to Kyllyngworth, levynge the cytie withoute ayde [p. 621].
[No matching passage]	And thus began rumour & malyce to spryng atwene the lordes of the lande: and specyally the duke of Somerset and other of the quenes counsayll were hadde in great hateryd fot the losynge of Normandy... For this yeldynge up of Normandy, moche dys-pleasure grewe unto the quene and hyr counceyll: in somoche that the duke of Yorke, father unto Edward the iiii. with many lordes with hym allyed, toke party agayn hyr and hyr counceyll, so that mortall warre therof ensuyd...[p. 626].

And when he come to the kyngis Tent the duke of Somerset was still awaytyng upon the kyng as chief abowte hym, and made the duke of yorke to Ride before the kyng thorwth london lyke a prysoner, and lyke As he shuld have been putt in holde, but tydyngis cam that therle of march sone was comyng with x M' wherfor he was libertid to goo at his wyll [p. 186].	And then was the duk of Yorke sent before to London, and was holden somedeale in maner as a prysoner; & more streyngther shulde have ben kepte, ne had ben tydynges which dayly sprange, yat syr Edwarde his sone, than erle of the Marche, was commynge towarde London with a stronge power of Welshemen & his Marchemen, which feyrd so the quene and hyr counceyll, that the duke was lybertyd to go where he wolde... [p. 627].

1452–53

And upon Seynt Edwardis day the Quene was delyvyrd of A ffayre prynce whos name was callid Edward [p. 186].	[the birth of a son to the king and queen] whose noble mother systeynyd not a little dysclaunder and obsequye of the common people, sayinge that he was not the naturall sone of kyng Henrye but changyd in the cradell, to hyr great dyshonour and heuynesse which I over passe...[p. 628].
[No matching passage]	Ye have in yeur remembrance how I before ye xxx. yere of this kyng, shewyde to you of the apoyntmente

	taken atwene the sayd kynge & duke of Yorke at Brent Heth, whiche apoyntment, as before is sayd was soone broken and set at nought, by reason whereof great enuye and discienuon grewe atwene the kynge and dyuerese of his lordys, and moost specyally atwene the quenys counceyll and ye duke of Yorke & his blode. For all contrary the kynges promyse, by meanys of the quene, which than bare ye cure & charge of the lande, the duke of Sumerset was set at large, & made capytayne of Calays...[p. 628].
1454–55	
[No matching passage]	...[the duke of York and the earls of Salisbury and Warwick collect their forces] Whereof the quene & the lordys were aduertyzed yat the duke was commyng with so great power, anon they caste in theyr myndes yat it was to none of theyr profetys...[p. 629].
sone aftir was A grete parliament by vertu where-of the duke of york made	...[following the account of the first battle of St. Albans] & all such

protectour of Engeland and the Erle of Warwyk, capitayn of Caleys And therle of Salysbury Chaunceler of Engeland, and all soche personys as had Rule before tyme abowte the kyng were voydid & put owte of Rule [p. 187].

persones as before were in auctoryte & nere about the kyng, were clerelye amoued & put by; and the quene and hyr cunceyll that before dayes rulyd, all utterlye set a parte concernynge the rule of the kynge and ye lande, whiche contynued for a whyle as after shall appere...[p. 629].

1454–55

Also in this yere the duke of yorke was sent for to Grenewych & there he was dyschargid of his protectourship

1456 grocer John Steward Thomas Canyngis Raaff Verney Anno xxxv

In this yere was therle of Salysbury dyschargid of Chauncelershyp [p. 189].

...The quene, with certeyn lordis which favoured hir partye, dysdayned sore the rule which the duke of Yorke bare and other, and speciallye for that that the sayde duke bare the name of protectour, which arguyd that the kynge was insuffycient to gouerne the realme, whiche, as she thoughte was a great dyshonoure to the kynge and to all the realm. Wherfore she made suche meanys, and wanne by hyr polycy such frendshyp of dyverse of the lordis, both spyrytuell and temporall, she causyd the duke of Yorke to be dyschargyd of his protectourshyp, and the erle of Salesbury of his chauncellershyp, which was cause of newe warre, as after shall appere...[p. 631].

1456–57

And aftir the duke of Yorke And the said Erle with ye Erle of warwyk were sent for by prevy seale to Cowentre whe they were alle nere be-trappid [p. 189].

This yere, and begynnynge of the same, the quene suspectynge the cytie of London and demyd it to be more fauourable unto the duke of Yorkys partye then hyrs, causyd the king to remoue from London unto Coventre, and there helde hym a longe season. In whiche tyme the duke of Yorke was sent for thyther by pryvey seale, with also the erle of Salesbury, & the erle of Warwyke, where, by couyne of the quene, they were all iii in great daunger...[p. 631].

1457–58

And the vijth day of march the kyng cam to london and the Quene at which tyme was made A Concorde and an unyte among these forsayd lordis, In token whereof upon owir lady day next ffolowyng which was the xxv day of march the kyng the Quene and these

...And sone affter, so to appease the rauncour and malyce atwene the quene and the other lordys, a day of matynge was appoyntyd by the kynge at London... [p. 632].

forsaid lordis went in procession Royally At seynt pawlis, And a noon aftir the kyng and alle these astatis departid [p. 190].

1458–59

In this yere aftir Candilmesse A man of therlis of warwyk smote A servaunt of the kyngis in Westmynstir halle, wherewith the kyngis howsold meyny were wroth and cam owte with wepyns and some for haste wt Spittis for to have slayn therle but the lordis that were his ffreendis conveyd hym to his barge, how be it many of his howsold meyny and wayters were hurt, And for this was grete labour made to the kyng to have hadde the sayd Erle Arestid but he Incontynently departid the towne toward warwyk, And sone aftir he purchassid A Commyssion and went to Caleys.... Also the same yere the Erle of Salysbury comyng toward london was Encountrid at blore heth wyth the lorde Audeley and his company which entendid to have takyn hym but he was before warnyd, and hadde in his company his ij Sonys sir Thomas & sir John Nevill, And a good

This yere about the feest of Candelmasse, the foresayd dyssymulyd loveday hangyng by a small threde, atwene ye quene and the for named lordis, expressyd in ye precedyng yere, the kynge and many lordes than beynge at Westmynster, a fray happenyd to fall atwene a servant of the kynges and a servant of the erlis of Warwyke...[the earl escapes]... For this ye olde rancur and malyce, which nevyr was clerely curyd, anon began to breke out; in somoch that the quenys counceyll wolde have had the sayde erle arestyd and commyttyd unto the Towre. Wherefore he shortly after departyd towarde Warrewyke, and by polycy purchased, soone after, a commyssyon of the kynge, and so yode or saylyd unto Calays.

ffelyshyp, at which metyng the lord Audeley was slayn and moche people of Chesshire, And the Erlis ij Sonys hurte, And goyng homeward they were takyn by the Quenys meyne and sent to Chester
Also in this yere was the ffeeld of ludlowe [pp. 190–91].

Than encreasyd this old malyce more and more, in somoche yat the quene and hir councyll sawe that they myght not be avengyd upon the erle, that so unto Calays was departyd, then they malygned agayne his father, the erle of Salysbury, & imagenyd how he myght be brought out of lyfe. And in processe of tyme after, as he was rydynge towards Salysbury, or, after some, from his lodgynge towarde London, the lorde Audeley, with a stronge company, was assygned to mete with hym, and as a prysoner to brynge hymn unto London. Wherof the sayd erle beynge warned, gatheryd unto hym the moo men, & kepyng his iourney, mette with the sayd lorde Audeley at a place called Blore Heth, where both companyes ran together & had there a stronge bykerynge; whereof in the ende theerle was vyctour, and slewe there the lord Awdeley and many of his retynew. At this skyrmysshe were the ii. sones of the sayd erle sore woundyd…; the which shortly after, as that were goynge homewarde, were by some of the quenys party taken…[the duke of York and his supporters raise forces and in October 1459] they drewe them towarde the kynge, to the entent to remove from hym suche persones as they thought were enemyes unto the common weale of Englande. But the quene and hyr counceyll herynge of the entent and strength of theyse lordys, causyd the kyng in all haste to sende forthe commyssyons to gather the people… [there follows the flight of the Yorkist lords at Ludlow]… [pp. 633–34].

It is obvious that the attribution of both *The Great Chronicle* and *The New Chronicles* to Fabyan requires reconsideration in the light of this evidence. It is tempting to argue that the differences between the two texts for the years 1450–59 amount, in effect, to a body of proof that he could not have been the author of both works. And this view is supported by an examination of *The New Chronicles'* account of the troubles of Sir Thomas Cooke: not only is this much more terse than that of *The Great Chronicle* but it contains no reference to the age of the author at the time or to his apprenticeship.[10] Even so, it is unwise to dismiss the consensus

10. *New Chronicles*, 656.

among later sixteenth-century authorities that Fabyan was the author of *The Great Chronicle*.

An assessment of the differences between the two narratives for the years 1450–59 can be reconciled with Fabyan's authorship of both if we assume that he wrote at least the portions dealing with these years at different times. And two considerations suggest, if this is the case, that *The New Chronicles* is the later of the two narratives. First, the references in it to the queen and her associates appear to be a level of interpretation, added to the basic sketch of events. Second, there is the fact that it is *The New Chronicles* which correctly connects the resignations of York from the protectorship and Salisbury from the chancellorship in 1455. This conclusion is, however, subject to one objection. The author of *The New Chronicles* stated that he completed this work in 1504, whereas *The Great Chronicle* was continued in the autumn of 1512, within a few months of Fabyan's death in 1513. It is, to be sure, possible to counter this objection by suggesting that, if Fabyan was the author of both narratives, he decided to continue a London chronicle after he completed *The New Chronicles*. But, if so, we have to concede that he did so in a manuscript that he had substantially revised in another version for the years 1450–59. This is not inconceivable. In the last resort, indeed, there are bound to be limits to our capacity to investigate Fabyan's putative authorship of both narratives, since we do not know how many copies were made, the relationship of the surviving manuscripts to those that have disappeared or, indeed, if the former are in Fabyan's own hand. In the light of this discussion it is difficult to reject the theory propounded by Thomas and Thornley, though it must be said that it appears much less convincing after an examination of the two narratives for the years 1450–59.

Even so, the differences revealed by such an examination of the entries for 1450–59 exist and must be explained. The explanation may lie in one or both of two areas. There is the possibility that after writing *The Great Chronicle* for the years 1450–59 Fabyan encountered another source for these years and chose to make substantial revisions when he came to write *The New Chronicles*. There is good reason to believe that this happened, at least to some extent. A comparison of both *The Great Chronicle* and *The New Chronicles* with the surviving continuation of *The Brut* for these years reveals its influence in one — *The New Chronicles*. It is also true that there may have been another major London chronicle which has disappeared; and, if so, this may be one used by a generation later by Edward Hall.[11] Is it possible, then, that this or another source contained the details relating to the activities of the queen and her associates that Fabyan incorporated in *The New Chronicles*? There is a compelling argument against this suggestion. Both *The Great Chronicle* and the relevant continuation of *The Brut* share with the other surviving contemporary chronicles a silence about the activities of Margaret of Anjou and her associates in the years prior to 1459 and an apparent unanimity that her active political role began in that year.[12] It is, indeed, possible to detect an echo of a lost source in the chronicle known as John Benet's when it tells us that in September 1456 "the king sent for the earl of Warwick and for the duke of York who came to the king and were very kindly received by the king though (*nam*) the queen hated them very much."[13] But the author does not mention any plot against them, as do three other chronicles,[14] let alone assign responsibility for it to the queen, and he otherwise attributes active intervention to her in the course of 1459.

A priori, the available evidence leads to another explanation: if Fabyan was the author of both *The Great Chronicle* and *The New Chronicles* he must have engaged in deliberate reinterpretation in the latter when he came to deal with the events of 1450–59. And, if the author of *The New Chronicles* was neither Fabyan nor the compiler of *The Great Chronicle,* the same conclusion that substantial reinterpretation occurred still holds, since there can be no doubt that *The Great Chronicle* was one of his main sources. Additional support for this conclusion is provided by two other considerations.

11. *Great Chronicle,* xviii, n. 3.
12. *A Chronicle of London from 1089 to 1483,* ed. N.H. Nicolas and E. Tyrrell (London, 1827), 135–40; *An English Chronicle,* ed. J.S. Davies, Camden Society, Series i, 64 (London, 1856), 66–85; *The Historical Collections of a London Citizen,* ed. J. Gairdner, Camden Society, New Series, 17 (London, 1876), 189–207 (*Gregory's Chronicle*); *Three Fifteenth Century Chronicles,* ed. J. Gairdner, Camden Society, New Series, 28 (London, 1880), 66–73; *The Brut,* ed. F.W.D. Brie, Early English Text Society, Original Series, 136 (London, 1908), 520–26; *Chronicles of London,* ed. C.L. Kingsford (Oxford, 1905), 158–69; *John Benet's Chronicle for the Years 1400 to 1462,* ed. G.L. Harriss and M.A. Harriss, Camden Fourth Series, 9 (Camden Miscellany, 24) (London, 1972), 198–224. All these were compiled in London or have London connections. But the St. Albans chronicler, who was in an especially good position to watch events, does not attribute an active political role to the queen until the Wakefield and St. Albans campaigns of 1460–61 (*Registrum Abbatiae Johannis Whethamstede,* ed. H.T. Riley, 2 vols., Rolls Series (London, 1872–73), 1:388 ff.
13. *Benet's Chronicle,* 217.
14. *Great Chronicle,* 189; *Brut,* 523; *Chronicles of London,* 167.

One is an argument in favor of the basic accuracy of the account of political events in *The Great Chronicle.* Sir Thomas Cooke had been the controller of Margaret of Anjou's household,[15] and had also acted as her agent in the export of wool.[16] It is likely that, if he had been aware of the political activities on the queen's part on the scale depicted in *The New Chronicles*, his reminiscenses would have been reflected, at least to some apparent extent, in the narrative written by his erstwhile apprentice, especially since the events that led to his troubles began with the exiled queen's approach for a loan. The second argument emerges from an examination of the relevant continuation of *The Brut.* In its description of the desertion of the Yorkist forces at Ludlow and the consequent predicament of the duke of York and the earls of Salisbury and Warwick it tells how they

> saw that the governaunce of the Reame stode moste by the Quene & hir counsell, and how the gret princes of the land wer called to Counsceil bot sett A-parte; & nat onely so, but that it was said thrugh the reame that the said lordes shold be destroyed utterly, as it openly was shewed atte Bloreheth bi thame that wold have slayn therl of Salesbury.[17]

It is true that some other contemporary chroniclers were more explicit in attributing responsibility for the Blore Heath episode to the queen. But it is *The Brut*'s language that *The New Chronicles* echoes in its account of 1459. And it was its phrase — "the Quene & hir counsell" — that probably supplied a recurring phrase in the interpretation of the events of 1450–59 that we find in *The New Chronicles.*

Why, then, did the writer of *The New Chronicles* engage in such reinterpretation? For the purpose of the remainder of this discussion it is sufficient to assume that he was Fabyan: he was his contemporary, if not he, and he was using *The Great Chronicle* and other sources that Fabyan would have used. In attempting to reconstruct Fabyan's mental processes we have no more evidence at our disposal than the texts themselves and the basic facts of

15. *Great Chronicle,* 205.
16. Public Record Office, London: Customs' Accounts, King's Rememberancer (E122), 73/26 (1449–50).
17. *Brut*, 526.

Fabyan's life. But it is reasonable to place these in the political context of the time. The starting point of an effort to do so has to be the contents and approach of *The New Chronicles* itself: it was explicitly described as a "concordance" of the chronicles of the two kingdoms of England and France. We know that Fabyan had read the *Compendium super francorum gestis* of the French author Robert Gaguin, published in 1497, since he cites this author.[18] He may well have wanted to provide a record of the relations of the two kingdoms from the English point of view. He belonged to a group of leading London merchants for whom the resumption of the French war meant enlarged profits through supplying the war effort and who saw benefits for their trade in an alliance with the ruling power in the Low Countries which would be an inevitable consequence of a French war. When Fabyan completed *The New Chronicles* the prospect of a French war must have seemed especially real, since Henry VII had moved into the framework of an alliance between the ruler of the Spanish kingdoms and the Hapsburg regime in the Low Countries. The second aspect of the background of the composition of *The New Chronicles* lay in the achievements of Henry VII within the English kingdom. Decades of civil war, the beginnings of which had been associated with disaster in France, had been brought to an end. Indeed, the period of writing must have coincided with the years of total security that Henry VII had enjoyed since 1497. It is true that Prince Arthur had died the year before the completion of *The New Chronicles*; but the younger son Henry survived as the living embodiment of the union of the two houses of Lancaster and York. To be sure, it was Edward Hall who incorporated the phrase "the union of the houses of Lancaster and York" into the title of his own chronicle a generation later; but his interpretation was already implied in *The New Chronicles* of Fabyan, ending in 1485.

It is hardly surprising that, writing in the England and the London of his time, and influenced by the experiences and memories of his early youth, Fabyan should have projected the political influence of Margaret of Anjou backwards from the events of 1459. It would be idle to speculate whether he was influenced at all by the Florentine historians who wrote in the fifteenth century;[19] but a comparison of *The New Chronicles* with *The Great Chronicle* and other English chronicles of the middle years of the fifteenth century does suggest

18. E.g., *New Chronicles,* 617, 620.
19. See e.g., D.J. Wilcox, *The Development of Florentine Humanist Historiography in the Fifteenth Century* (Cambridge, MA, 1969).

that Fabyan had a much more developed historical sense than these authors and, in particular, a greater concern with the causes of the events they all narrated, even though he clung to a system of dating by mayoral years. His early adolescence began shortly after an army led by the queen, notorious because of its depredations during its march south, threatened to attack London after the second battle of St. Albans in February 1461; and the belief that the queen disliked the Londoners can be detected in the autumn of 1456.[20] The picture of the queen as the determined leader of the Lancastrian cause could only have been strengthened by Yorkist propaganda during the years her supporters in the north of England were being defeated.[21] For Fabyan to blame her husband Henry VI for the rise of faction and the outbreak of civil war was unthinkable; indeed, he wrote during years when Henry VII was seeking his canonization and the transference of his relics to Westminster Abbey.[22] There could be no doubt that the queen became the leader of the Lancastrian cause and protagonist of her son's right of succession, at least in 1459. It was inevitable that a thoughtful writer of Fabyan's time would ponder her role in earlier events; and her French nationality made her an especially convenient scapegoat against the background of centuries-old antagonism of the kingdoms of England and France. There is good reason to believe that an interpretation that extended the queen's active role in politics backwards to 1450 was not created by Fabyan alone. The Flemish chronicler Waurin, who was in England and in touch with court circles in 1467 and completed his chronicle in 1471 or shortly after,[23] presented an account that resembled Fabyan's, though it contained clearly fictitious details, and unlike Fabyan's, depicted the queen as influencing the king in favor of the duke of Suffolk.[24] These details suggest that Margaret of

20. *Benet's Chronicle,* 217.
21. *Letters of the Kings of England,* ed. J.O. Halliwell, 2 vols. (London, 1848), 1123–30. For example, the first letter referred to the alliance the ex-queen had made with "Scots and Frenchmen, whom she hath excited and provoked to show them of the greatest cruelty and tyranny against our said subjects that they can, unto the execution of the end of her insatiable malice towards them: wherewith her joy and consolation is most disposed and applied."
22. B. Wolffe, *Henry VI* (London, 1981), 355–58.
23. Jehan de Waurin, *Recueil des croniques et anchiennes istories de la Grant Bretaigne a present nomme Engleterre,* ed. W. Hardy and E.L.C. Hardy, Rolls Series, 5 vols. (London, 1864–91), 1:xlii, cxciii–cxciv.
24. Ibid., 4:363, 5:255.

Anjou was viewed as responsible for England's troubles as early as 1467.[25]

It has long been recognized that the Tudor historical tradition, which at the end of the Elizabethan period found final expression in Shakespeare's plays on English history, had its earliest origins in Fabyan's *The New Chronicles*.[26] It has also been stated by the standard authorities[27] that Fabyan's account of fifteenth-century England was used by Polydore Vergil who in turn influenced Edward Hall in the writing of his *Chronicle*. Even on the basis of the dates of composition and publication, however, there are serious difficulties in deriving Vergil's account from Fabyan's. Fabyan completed *The New Chronicles* in 1504. It was not until 1506 that Vergil was asked to write a history of England by Henry VII;[28] and we must assume that he did not reach the fifteenth century until a short time before he visited Italy in 1514–15, when he took with him the manuscript of the first redaction of his *Anglica historia*.[29] And this did not appear in print until 1534, whereas *The New Chronicles* was published by Pynson in 1516.

An examination of the manuscript[30] of the first redaction (in Vergil's own hand) reveals a much different relationship between Fabyan's account and that of Vergil. There are marked variations between the way in which Vergil handled the queen's role in the manuscript and in the two editions printed in his lifetime. In all of these he stressed the queen's hostility towards the duke of Gloucester directly after her arrival in England and described how, egged on by her father, the king of Sicily, she determined to bring about the end of his authority and power in the kingdom.[31] But in the manuscript she was assigned no

25. According to the French chronicler, Mathieu d'Escouchy, the queen was seeking French help as early as 1457 and was responsible for the French attack on Sandwich in that year. See M. d'Escouchy, *Chronique*, ed. G. du Fresne de Beaucourt, 3 vols. (Paris, 1863–64), 2:352–54. His chronicle was completed in 1465 (ibid., 1:xxxix).
26. Gransden, *Historical Writing,* 245.
27. Cf. C.L. Kingsford, *English Historical Writing in the Fifteenth Century* (Oxford, 1913), 254–55; D. Hay, *Polydore Vergil, Renaissance Historian and Man of Letters* (Oxford, 1952), 87, followed by Gransden, *Historical Writing,* 442. Hay simply quotes Kingsford *in extenso*.
28. Hay, *Polydore Vergil,* 79.
29. Hay concludes that the manuscript was written in 1512/13. See D. Hay, "The Manuscript of Polydore Vergil's *Anglica Historia,*" *English Historical Review* 54 (1939): 243.
30. Vatican City, Biblioteca Apostolica Vaticana, Cod. Lat. Urbino 497, 498.
31. Ibid., 498, f. 174. The passage is, however, markedly more concise than those in the printed editions. This fact and the details that follow are not mentioned in Hay's collation of the manuscript with the printed versions (*Polydore Vergil,* 194–95).

role in the exile of the duke of Suffolk,[32] in the handling of the duke of York in 1450–52,[33] in the plot to inveigle York and the Neville earls to Coventry[34] and the attempted assassination of the earl of Warwick, this last episode being explicitly attributed to *regis familiaribus*.[35] If the manuscript version is treated in isolation from its printed counterparts, it is quite clear that Vergil was writing the portion dealing with the years 1445–61 under two main influences — the version of *The Brut* printed by William Caxton in 1480 and at least one account of the queen's attitude to Gloucester. Vergil's manuscript is the earliest source of English provenance to mention this aspect of the queen's role; and it is impossible to say whether he had access to a written account that has since disappeared or simply absorbed stories that were current in court circles in the years following his arrival in England in 1502.[36] At any rate, the details in Vergil's manuscript show that the myth of "the good duke Humphrey" must have been current by the time he was writing his first version of the *Anglica historia*. It is tempting to see in one detail an indication that Vergil was still in the process of developing his view of the queen's role in the course of writing his manuscript. In the passage immediately preceding that in which he described the birth of a son to the queen and her husband, he described how the duke of Somerset controlled the kingdom's government but added the phrase *una cum Margareta regina* by means of a marginal insertion.[37]

These details, viewed as a whole, show that a process of reinterpreting the queen's role in English politics was developing in the mind of Polydore Vergil just as it had a decade or two earlier in that of Fabyan. A comparison of the first printed edition of 1534 with *The New Chronicles* reveals that Vergil had now decided to

32. Vat. Lib., Cod. Lat. Urb. 498, f. 178v. The decision is attributed to the king alone. It should, however, be noted that in a preceding passage narrating the period between the death of Gloucester and the loss of Normandy Vergil describes the king's unworldly demeanor, which he contrasts with the queen's dependence on bad counsel and inability to control the situation (ibid., f. 176).
33. Ibid., f. 180v.
34. Ibid., f. 184v. The king's move to Coventry was, however, attributed to Somerset and other nobles in cooperation with the queen.
35. Ibid., f. 185.
36. It is obvious that we have to dismiss Hay's suggestion (*Polydore Vergil,* 93) that he obtained information from Cardinal Morton, since the latter died in 1500 and Vergil did not arrive in England until 1502.
37. Vat. Lib., Cod. Lat. Urb. 498, f. 181.

stress the queen's role in most of those episodes in which Fabyan thought her an active force and in which no such role had been attributed to her in his manuscript version.[38] Accordingly, it is the printed editions of the *Anglica historia*, as distinct from the original version, which were influenced by Fabyan's *The New Chronicles*.

The respective dates of publication leave little doubt that the *Chronicle* of Edward Hall was influenced by Fabyan's *New Chronicles*. Nevertheless, there is a problem in assuming that he was also influenced by Polydore Vergil. Hall's *Chronicle* was largely completed by 1532[39] and Vergil's *Anglica historia* was not published until 1534 (at Basel). Hall did not die until 1547[40]; and there is some reason to believe that revisions and additions, albeit not in the form of final drafts,[41] occurred in the intervening years. If Hall did incorporate elements of Vergil's account, these must have been accretions to a text already indebted to Fabyan. Certainly there are strong resemblances between Vergil's details and the narrative of Hall in the case of Margaret of Anjou's activities in the years 1445–50, Hall developing the queen's complicity in the murder of Gloucester and attributing to her an affair with Suffolk.[42] But it is also possible that Hall's picture of the queen's involvement in the faction struggles that led to the Yorkist seizure of the throne was influenced by a body of opinion that had enlarged and deepened her role in the years that had intervened since Polydore Vergil had written his first redaction of the *Anglica historia* over twenty years before.

The historian who studies the "Wars of the Roses" has to exercise a careful judgment about the extent to which one can ascribe motives and stress the role of individuals outside the limits of the information supplied by contemporaries. There are grounds for believing that the queen did seek to obtain the regency in January-

38. *Three Books of Polydore Vergil's English History Comprising the Reigns of Henry VI, Edward IV and Richard III from an Early Translation,* Camden Society, Original Series 29 (London, 1844), 77, 79. This, a mid-sixteenth-century translation of the edition of 1546, is the most convenient means of access to the printed version. For its account of the queen's role in the exile of Suffolk (which states that she "could well spare him out of her sight"), see ibid., 83.

39. A.F. Pollard, "Edward Hall's Will and Chronicle," *Bulletin of the Institute of Historical Research* 9 (1932): 173–74.

40. Ibid., 171.

41. Ibid., 172–73.

42. E. Hall, *Chronicle*, ed. Sir Henry Ellis (London, 1809), 208–9, 218–19.

February 1454 during her husband's mental illness.[43] It is also reasonable to infer that the birth of her son increased her suspicion of York's intentions and made her determined to protect her child's right of succession to his father. Yet the earliest contemporary evidence that the queen was acting against York comes from her effort to obtain the regency. And, apart from a few letters in the Paston collection which may only reflect rumor and gossip, and terse comments by Thomas Gascoigne,[44] there is little solid information to support the view that the queen took over the leadership of the Lancastrian cause in 1456.

For historians of earlier generations the robust Tudor tradition, assigning political leadership from her arrival in England onward to a queen who was married to a political simpleton, had a compelling quality. Even T.F. Tout in his article on Margaret of Anjou in the *Dictionary of National Biography* treated the Tudor chronicles as authentic evidence. In contrast with this we can place the two most recent scholarly studies of the reign of Henry VI which ignore the Tudor tradition and are cautious in their handling of contemporary sources.[45] Yet echoes of the interpretation of the queen's role that can be traced back to Fabyan can still be detected in other recent writings.[46] The historiography of the beginnings of the "Wars of the Roses" has not totally escaped from the shadow of *The New Chronicles* of Robert Fabyan.

■

43. For a useful and cautious discussion, see R.A. Griffiths, *The Reign of Henry VI* (Berkeley and Los Angeles, CA, and London, 1981), 722.
44. Thomas Gascoigne, *Loci e libroveritatum,* ed. J.E.T. Rogers (Oxford, 1881), 214.
45. Griffiths, *Henry VI,* esp. 772, 775–77; Wolffe, *Henry VI,* esp. 276, 302–3.
46. E.g., R. L. Storey, *The End of the House of Lancaster* (London, 1966), where the concluding chapter is entitled "The King's Peace and the Queen's War."

XIII
Sir Thomas More and the Greek Anthology

ALAN CAMERON

THE GREEK ANTHOLOGY is one of the great books of European literature. It has been described as a garden containing the flowers and weeds of 1500 years of Greek poetry. It is a book that has fascinated me ever since I first read a few poems from it at St. Paul's School in London some thirty years ago.

It is not (of course) a book so much as a collection.[1] The word "anthology" means a collection of flowers. Indeed the word was probably coined for a collection of epigrams; at any rate it is first found so applied in the second century C.E. The first Greek Anthology was put together by Meleager of Gadara c. 100 B.C.E., comprising his own work together with a substantial selection from the epigrams of his predecessors, mostly of hellenistic date, but a few going back to the sixth century B.C.E. He called it his *Garland*, and in a long preface works out the metaphor in detail, identifying every poet he includes with a flower and weaving them all together. The *Garland* itself has not survived, though a couple of papyrus fragments have recently been published. He had many successors, including Philip of Thessalonica, who put together a second *Garland* in the reign of Nero, and Strato of Lampsacus, who wrote pederastic poems under Hadrian. The last was Agathias, a lawyer writing in the very different and now Christian age of Justinian at Constantinople. There is thus an amazing range, chronological as well as thematic, touching on every imaginable subject, from the highest poetry to the merest doggerel.

All these ancient collections perished, and we would know practically nothing about them and hardly more about most of the poets they anthologized but for one man. His name was Constantine

1. The standard edition is now that of H. Beckby, *Anthologia Graeca*, 4 vols., 2nd ed. (Munich 1967–68). The main facts and problems are clearly set out by A.S.F. Gow, *The Greek Anthology: Sources and Ascriptions,* Hellenic Society Supplementary Papers 9 (London, 1958). For further details and some new solutions I refer to my own forthcoming study, *The Greek Anthology: From Meleager to Planudes* (Oxford, 1987).

Cephalas ("Bighead"), and it is to him that we owe the Greek Anthology we know. About the year 900 this Constantinopolitan schoolmaster made a huge collection of his own based on all the earlier collections he could find. Fortunately the epigram had been in high vogue at the court of Justinian, and copies of Meleager, Philip, Strato, Agathias and others were still easily found in the libraries of Constantinople.

Cephalas clearly drew on copies of all these collections. To some extent he followed Meleager's own example and tried to construct a thematic arrangement of his own, but after 200 or 300 poems in each of his categories his energy usually gave out, and he ended by just copying out chunks of 100 poems or so from each of his major sources in turn.

This is particularly clear in the case of Philip, since Philip's own principle of arrangement unmistakably shines through. Philip had divided up his poems according to the first letter of the first word of each poem. Thus all poems beginning with an alpha come before all poems beginning with a beta, and so on. Now from time to time in Cephalas we run across a sequence of up to 100 poems in pure alphabetical sequence — poems by writers known to have contributed to Philip's *Garland* (his preface lists his contributors). One of these sequences starts at omega and works backwards for a change.

That much has long been obvious, but I have noticed an interesting further detail. Some of Cephalas' main books, on erotic and funerary poetry, have successive sequences from Meleager and Philip, then some different arrangement, and then more sequences from Meleager and Philip. It might appear that he was simply breaking his material up for the sake of variety. But there are two curious features that point to a different conclusion. In the second of these sequences there is usually at least one duplication, poems already copied in the first sequence. And in every case of a duplicated poem, the two occurrences present a different text. The simplest explanation is that Cephalas had *two* copies of the *Garlands* of Meleager and Philip, both no doubt somewhat reduced versions of the originals. He copied out the erotic or funerary poems from the first copy in his first sequence, and then added from his second copy material missing from the first, together with those occasional duplications.

Grateful though we are to Cephalas for his priceless compilation, there is no denying that he was a hasty and mechanical worker. Paradoxically enough, this turns out to be one of the most valuable features of his collection.

For example, it seems clear that Meleager did not himself distinguish between hetero- and homosexual epigrams in his *Garland*. Rather he had one continuous sequence of erotic poems arranged by theme, irrespective of the sex of the beloved. This was to be expected from a man like Meleager, whose own love poems are addressed, with admirable impartiality, to eight different girls and eleven different boys. Cephalas decided to have separate books for hetero- and homosexual poems. This was not a wise decision. It might not have seemed too hard to distinguish between them — but only for someone who was prepared to take the trouble to read the poems carefully, and Cephalas never had time for that. In the first place, he did not realize that there was a large category of Greek feminine names ending in *-on*; all such ladies he transferred lock, stock and barrel to his pederastic book. Then there are cases like a self-apostrophe by the poet Asclepiades beginning "Drink up, Asclepiades...." Only reading as far as the masculine apostrophe, Cephalas at once popped this piece too in with the pederastica. It is precisely the mess Cephalas made of his two erotic books that tells us that Meleager made no such distinction.

Perhaps the single most absurd blunder Cephalas made is in a section of his own compilation on ancient cities. The poem in question celebrates Heraclea (*AP* 9.554). Now Heraclea is a common name for Greek cities, though it is also found, more rarely, as a name for women. Cephalas evidently read no further than the name in line 1 and the word *polis*, city, at the end of line 2. If he had read on he would have discovered that it was a poem on a girl called Heraclea said to have been the best fellatrix in the city.

It would be hard to exaggerate the significance of Cephalas' Anthology. On the one hand he preserved for us several thousand epigrams, some of the highest quality, that we would never otherwise have known even existed, from the sixth century B.C.E. to the sixth century C.E. Not only that. He took great pains to ascribe every poem he could to its author, however obscure. Thanks to Cephalas we know of hundreds of poets otherwise unknown. On the other hand, thanks to his mechanical methods of arrangement, his haste and his blunders, we are better able to track down his sources than if he had done the job properly. But for Cephalas we should have known absolutely nothing of the anthologies of Meleager, Philip, and Agathias. Thanks to him, we can to a large extent reconstruct them.

Now anthologies are not original compositions, the sole patrimony of their compilers, and they tend not to be treated as protected texts. They are abridged and selected, bowdlerized, expanded with new material that has attracted a later owner. And this is even more obviously true in an age when every line of every text had to be copied by hand. Faced with ten epitaphs on grasshoppers, a man might drop one or two of them, sometimes deliberately, sometimes inadvertently — or add a new one of his own gathering. The two papyri we have of Meleager are clearly fragments from a drastically abbreviated text.

Cephalas' claim on his material was even less than (say) Meleager or Philip or Agathias, since he was not a major contributor to his own anthology. It is not surprising that it has not survived entire and intact under his own name. We have two witnesses, of very different age and authority: the Palatine Anthology, of the early tenth century, and the Planudean Anthology, of the early thirteenth.

First the Palatine Anthology, from the Palatine Library in Heidelberg, palatinus gr. 23, generally known as *AP*.[2] Having *AP* is almost as good as having Cephalas' own manuscript. It dates from c. 940, less than half a century after Cephalas, and as we learn from a series of invaluable marginalia by a contemporary owner, it was corrected against the copy of one Michael the archivist. This copy of Michael's, we are told, was made directly from Cephalas' autograph. The margins are full of corrections entered from Michael's manuscript, in particular authors' names omitted by the original scribes of *AP*. The man who compiled *AP* made some additions of his own, mainly Byzantine material of less interest. Because of its age, contents and informative marginalia, *AP* is one of the most important and interesting of all surviving Greek manuscripts. It casts a flood of light on scholarship in the age of its compilation.

2. The fullest description is K. Preisendanz's introduction to *Anthologia Palatina: codex Palatinus et codex Parisinus phototypice editi* (Leiden, 1911) — a classic product of the *Quellenforschung* of its age, an indispensable collection of facts but coming to preposterous conclusions.

It was one of the more exciting moments in my research when, almost accidentally, I hit on the identity of the man responsible for *AP*: the compiler, editor and in part scribe of the manuscript, hitherto known only by the letter J. In a part of the manuscript written in the hand of J there is a small group of contemporary epigrams, poems dating from the early tenth century. One is ascribed to Constantine the Rhodian (*AP* xv.15), a poet known from elsewhere to have been alive as late as 950. That is to say, he may still have been alive when *AP* (of c. 940) was written — certainly the latest identifiable contributor to *AP*. Now this poem is ascribed to "the humble (*tou tapeinou*) Constantine the Rhodian." This self-depreciatory formula is in fact one of the commonest formulas of self-ascription in the Byzantine period. This does not in itself prove that the scribe J *is* Constantine the Rhodian. He might simply have been copying out a poem from Constantine's own copy. But there is a clincher. The last two lines of the poem are clearly alternative drafts; only one of them is needed to complete the poem, and the first of the two is horribly unmetrical. Anyone who looks at the manuscript (which can be done easily enough in Preisendanz's magnificent facsimile), will see a detail not reported in printed editions: there is a deletion mark in the margin against the less metrical of the two alternatives. That is to say, it was already marked for deletion by J. Nobody writes alternative lines for someone else's poems. This is Constantine himself, improving his own poem in his own hand.

So MUCH FOR *AP*. The other witness to Cephalas is the Anthology of the Byzantine monk Maximus Planudes. Here we are lucky enough to possess Planudes' own autograph (Venice, Bibl. Naz. Marciana, MS gr. 481), dated in his own hand to September 1301.[3] Planudes' Anthology is thus 350 years later than *AP*, drastically reduced in size, and in most respects vastly inferior. In fact there is only one respect in which it might be said to excel Cephalas, and that is in its arrangement. Planudes was nothing if not methodical and industrious, and he completely reorganized Cephalas' material into countless thematic categories. So well did he do the job, in fact, that he obliterated all trace of the original arrangement of Meleager, Philip, and Agathias. He never mentions their anthologies — or

3. Alexander Turyn, *Dated Greek Manuscripts of the Thirteenth and Fourteenth Centuries in the Libraries of Italy* (Urbana, 1972), 1:90–96; and in *Epeteris Etaireias Buzantinon Spoudon* (1973), 403–50.

Cephalas either. But for *AP*, we should never have known that they existed.

The Planudean Anthology also suffered heavily from the blue pencil of the man who has been dubbed the Dr. Bowdler of Byzantium. Not only did Planudes simply omit large numbers of the erotic and pederastic poems included in such profusion by Cephalas. He deleted passages from poems he did include and often rewrote poems himself. Unlike Cephalas, he had a good understanding of classical metrics and was quite capable of substituting phrases of his own composition in passages he found objectionable. In a poem of Callimachus that refers to "love of a man" he neatly substituted the words "love of another woman"; in a poem where the writer admires "beautiful buttocks," he is made instead to admire "beautiful girls." One steamily carnal pederastic poem seems to have caused the good monk to lose his head, since for once he substituted an unmetrical word for "girl" where there was at least one obvious metrical alternative. But in general he covered his tracks neatly enough, and but for *AP* we would not have been able to reconstruct the poems he tampered with.

Perhaps because it was smaller and better arranged, perhaps because of changing literary taste, Planudes' Anthology was much more successful than Cephalas'. A large number of mostly unabridged and exact copies survive in manuscript — and under Planudes' name. And several printed editions followed the *editio princeps* of 1494. Two remarkable volumes by James Hutton trace the enormous popularity and influence of *the* Greek Anthology from the Renaissance down to 1800.[4] In fact till 1800 the Planudean Anthology was *the* Greek Anthology, the only Byzantine bestseller.

This is the question to which I want to address the rest of this paper. Why, in an age when every educated man had a copy of the Greek Anthology on his coffee table, when scholars everywhere read, quoted and translated the epigrams it contained; why did no one recognize and publish *AP*, an infinitely superior work, 350 years older, with a better text, all that information about its sources, and 1000 or more additional epigrams? Janus Lascaris, in publishing the *editio princeps* of the Planudean Anthology, had already spoken slightingly of Planudes' editorial work in general and bowdlerization (of which Planudes himself had boasted) in particular. So readers

4. *The Greek Anthology in Italy to the Year 1800* (Ithaca, NY, 1935); *The Greek Anthology in France and in the Latin Writers of the Netherlands to the Year 1800* (Ithaca, NY, 1946; rev. ed., 1967).

were prepared for the possibility of a fuller (and naughtier) Anthology.

Yet *AP* was not in fact discovered till 1606, and by a remarkable chain of circumstances not finally published till 1800. It was apparently a young scholar called Salmasius who first appreciated the value of the manuscript where most of it (after a brief exile in Paris) is still to be found today, the Palatine Library of Heidelberg. His superior, the librarian Gruter, at once sent a transcript of a dozen or so of the new naughty poems to Joseph Scaliger, the greatest scholar in Europe. Scaliger wrote back in great excitement[5] that this had to be the manuscript his friend Francesco Porto had reported seeing in the possession of Nicholas Sophianos. On the basis of this letter of Scaliger it has been concluded that *AP* was in Italy throughout the sixteenth century, in the possession of Porto and Sophianos, a Greek from Corfu who collected manuscripts, published a map of Greece and lived in Rome and Venice.[6] There is also evidence that the Cretan scholar Musurus knew *AP* in Venice early in the sixteenth century.[7] As a result, it has generally been assumed that *AP* was in Italy between at any rate 1500 and 1600, and then somehow got to Heidelberg. Since Francesco Porto's son Emilio happened to be living in Heidelberg when *AP* was discovered, not surprisingly it has been inferred that it passed down from Francesco to Emilio, who took it to Heidelberg with him and then gave or sold it to the Palatine Library.

Unfortunately this neat scenario falls apart on inspection. In the first place Scaliger does not say that Porto owned the manuscript that contained those naughty poems. On the contrary, Scaliger's words suggest a manuscript that Porto did *not* own, a possession of this man Sophianos that Porto coveted. Second, the idea that it was Emilio Porto who took *AP* with him to Heidelberg can be disposed of once for all. There is in the Bodleian Library a manuscript commentary on the Anthology by Emilio Porto, in his own hand, dated only two years before the discovery of *AP* in Heidelberg.[8] The Bodleian possesses many other manuscripts and books of Emilio Porto,

5. *Iosephi Scaligeri Epistulae* (Leiden, 1627), 798, Ep. 430.
6. For further information on all these figures, see my forthcoming *Greek Anthology* (n. 1 above).
7. E. Mioni, "L'Antologia Graeca da Massimo Planude a Marco Musuro," *Scritti in onore di Carlo Diano* (Bologna, 1975), 296–307.
8. F. Madan, et al., *A Summary Catalogue of Western Manuscripts in the Bodleian Library* (Oxford, 1897), 4:101.

and there can be no doubt about the identity of the hand. I have examined the book, and it is certain that Emilio had never seen or heard of *AP*. His commentary is entirely based on the Planudean Anthology.

As for the manuscript that Scaliger in 1607 claimed to have heard of at two removes, how can we possibly claim that it was *AP* simply because it had a few naughty poems not in the Planudean Anthology? There survive two other, much smaller collections that preserve a few of the erotic epigrams omitted by Planudes, most of which (naturally enough) also appear in *AP*. Porto's manuscript may well have been one of the three extant manuscripts of the so-called *Appendix Barberino-Vaticana*, at least two of which were in Italy during the Renaissance.[9] This collection contained 50 Palatine erotica, and if by some chance Scaliger had identified some phrase or theme in Gruter's transcript from Porto's description, that would still not prove that the manuscript Porto saw was *AP* itself. In any event, in 1607 it was nearly 40 years since Scaliger had met Porto in Geneva, and a further 20 since Porto knew Sophianos, probably in Ferrara or Venice. It might be added that (remarkably enough) Scaliger confused two different Greek scholars of the age called Sophianos, Nicholas and Michael. It will be clear that we can place little reliance on his recollection; after all, he never saw the poems himself.

I suggest that we forget all about Scaliger's confused and excited letter, and turn instead to four other documents that seem never to have been put together. They will lead us to the perhaps unexpected conclusion that during most of the sixteenth century *AP* was not in Italy at all, but (of all places) in England.

First there is Henricus Stephanus' first edition of the *Anacreontea*, published in 1554. The only manuscript that contains this extraordinarily influential little text is *AP*,[10] and Stephanus' edition agrees to the letter with *AP*. Unfortunately, Stephanus does not say what manuscript he used or where, but in his edition of the Planudean Anthology in 1566 he refers to a "vetus codex epigrammatum" from which he quotes a few riddle epigrams which, again, are only to be found in *AP*. He adds that

9. E. Mioni, *Miscellanea* 1 (1978): 69–79.

10. See now the new edition by M.L. West, *Carmina Anacreontea* (Leipzig, 1984).

this codex was in the possession of an Englishman called John Clement at Louvain when he saw it.

John Clement was a practising doctor who also edited Galen. He was at Louvain between 1549 and 1554, a Catholic in exile from protestant England in the reign of King Edward VI. Stephanus was there in 1551. It must have been in 1551 that they met.

The second text is an inventory of Clement's books prepared by his son in 1572, the year of his father's death, still in exile.[11] Among his manuscripts is listed an "epigrammatum liber magnus et perantiquus" that contained "twice as many epigrams as the Aldine Anthology and many more even than Stephanus' Anthology." Now Stephanus' edition of the Planudean Anthology (1566) was enriched by much additional material, and was by far the largest Greek Anthology then available.

Third, there is a note written by Gerard Falkenburg of Nymegen, editor of Nonnus' *Dionysiaca* (1569), on the title page of his copy of the Juntine Planudean Anthology[12] (1519) c. 1566 (the date he wrote his name on the flyleaf):

> Huius florilegii collector fuit Planudes, qui, e vetere circulo, epigrammata obscoeniora sustulit, quae etiamnum hodie apud Io. Clementem Britannum exstant.

The book is full of marginalia in Falkenburg's hand, but no new poems or readings from *AP*. Indeed his claim that Clement's manuscript contained merely the "naughtier" poems that Planudes had omitted proves that he had no idea of the true extent of its superiority. But the little he did know of Clement's Anthology is accurate as far as it goes, and certainly confirms that it was *AP*. A "very old" anthology of epigrams that was larger than Stephanus' and contained in addition riddle poems, the *Anacreontea* and numerous erotic poems omitted by Planudes simply has to have been *AP*.

Clement's family evidently fell on hard times towards the end of the sixteenth century, and it must have been then that the manuscript was sold, passing into the possession of the Palatine Library. The exact date cannot be determined; probably not before 1584, since F. Syllburg's catalogue compiled in that year makes no mention of it.

11. Published by G. Mercati, "Sopra Giovanni Clement e i suoi manoscritti," *La Bibliofilia* 28 (1926): 81–89 (= *Opere Minori,* Vatican 1937, 4:292–315).
12. Now in the Bodleian Library, Auct. S.5.33.

That *AP* was in Clement's possession between 1550 and c.1572 we may now regard as certain. But where did he get it? Clement was a protégé of Sir Thomas More; in fact he lived for a while in More's house and married his favorite niece (he is mentioned in the preface to *Utopia*). He was also the first identifiable pupil at St. Paul's School, founded by Colet with the backing of Erasmus and More in 1509. In 1520 he went to Louvain to study medicine, and between 1522 and 1525 he studied at Siena. According to V. Rose,[13] it was during his travels that Clement acquired *AP*, perhaps presenting it to More on his return to London. Clement did pick up a few manuscripts on his travels, but nothing else of any real antiquity or value. His only biographer has questioned the likelihood of a medical student having either the connections or the resources to make such an acquisition.[14]

Now More himself was keenly interested in classical Greek epigram and in 1518 published some Latin translations from the Anthology jointly with William Lily, first high master of St. Paul's School.[15] We know from Erasmus that most of these translations had been made many (some as many as 20) years before 1518.[16] But it is likely that at least one or two were more recent, perhaps made specially for the printed edition. Almost all the originals come from the Planudean Anthology, but one, no.18, does not. Since More and Lily actually quote the Greek originals, there can be no doubt about the identification of the poem they translate: Simonides *Epigram* 186 Bergk = 55 Page. This poem is transmitted in one very brief anthology preserved in three very late[17] manuscripts which almost certainly never left Italy in More's lifetime — and on the very first page of the Palatine Anthology. Under other circumstances one might have been prepared to postulate another, lost source. But since More's disciple Clement undoubtedly owned *AP* in the years after More's death, it hardly seems outrageous to suggest that

13. *Anacreontea* (Leipzig, 1876), vii.

14. E. Wenkebach, *John Clement: ein englischer Humanist und Arzt des sechzehnten Jahrhunderts,* Studien zur Geschichte der Medizin 14 (Leipzig, 1925), 64, n. 73.

15. See now the elaborate new edition by C.H. Miller, L. Bradner, C.A. Lynch and R.P. Oliver, *Latin Poems,* Complete Works of St. Thomas More 3.2 (New Haven, 1984).

16. Miller et al., 10–11.

17. Two written by Bartolomeo Comparini da Prato at Florence in 1493; the third by Scipione Carteromacho no earlier than this: R. Aubreton, *Revue des Études Anciennes* 70 (1968): 70–72; E. Mioni, "L'Antologia," 278.

More himself may have owned it a few years earlier in London. If so, how did he come by it?

Not through the offices of Clement, it would seem. For Clement did not leave England until 1518, and did not return until 1525. There is in fact another possibility.

There is solid evidence that the manuscript was (at any rate briefly) in Italy around 1500. Musurus gave lectures on the Anthology in 1506, and two extant manuscripts, at least 50 years later in date, preserve notes on the Anthology by Musurus. These notes quote a number of epigrams only preserved in *AP*, usually complete with the textual errors of the *AP* version. They also add a great many ascriptions of authors' names missing in Planudes but present in *AP*.[18] It is difficult to doubt that Musurus had made a close study of *AP*. But where and when?

It has hitherto been assumed, on inadequate grounds and in ignorance of the evidence about Clement and More, that *AP* was in Rome at this period, and that it was there Musurus read it. But Musurus did not visit Rome till 1516 and died in 1517. More's epigrams were published in London in 1518. One or two scholars in More's circle visited Rome around the turn of the sixteenth century. Lily himself in 1489 as a youth of 20; Colet (the founder of St. Paul's School) briefly in 1493; William Latimer in 1498-1505 and again in 1511.[19] But no one who could have brought More *AP* between 1517 and 1518. In any case, there is no evidence whatever that *AP* was ever in Rome, much less that Musurus saw it there.

The place where Musurus is known to have lectured on the Anthology is at Padua, in 1506. It would seem most natural to assign his notes on the Anthology and his consultation of *AP* to this date and place. Now this happens to have been the very date and place where More's friend Erasmus got to know Musurus. One of his letters records a visit to Musurus' house for dinner one day in November 1508.[20] In December of that year Erasmus left Padua for a brief stay in Rome, and then, early in 1509, to England — and More.

It is obviously quite possible that Musurus showed *AP* to Erasmus during his stay in Padua. That he *gave* it to him is less likely.But then there is no reason to suppose that Musurus actually owned it. It might have been in the possession of some private collector,

18. E. Mioni, Ibid., 290–307.

19. G.B. Parks, *The English Traveler to Italy* (Rome, 1954), 1:464–68.

20. *Opus epistolarum Des. Erasmi,* ed. P.S. Allen and H.M. Allen (Oxford, 1924), 5:244.

perhaps unaware of its value,[21] who took a fancy to Erasmus and gave — or perhaps simply lent — it to him.[22] We learn from the reports of scholars like Manutius, Musurus and Politian of many a valuable Greek codex in private hands that has now disappeared.[23] *AP* has at least survived. If he saw a chance of acquiring such a treasure, Erasmus would surely have seized it. The surprise is rather that, having acquired it, he gave it to More. But then we know that he had a deep admiration and affection for More.

THE HYPOTHESIS that *AP* was in London and the Low Countries between 1509 and 1572 would kill two birds with one stone. It would explain how early Renaissance scholars in Italy knew *something* of *AP* (Musurus' notes); but also why they knew no more — above all why no one published so obviously important a text. It was far away, inaccessible in Britain.

So the Palatine Anthology may be the most celebrated Greek manuscript ever to find its way to Britain — and then get away again, a victim of religious persecution. It is (of course) rather satisfying for a former pupil to find this wonderful manuscript already at St. Paul's School 450 years before he learned of its existence there. But at the same time more sober reflection suggests that if it had been allowed to stay in Venice or Padua, it would surely have been published within a few years, and the subsequent history of the Greek Anthology would have been very different.

■

21. Given the entirely different arrangement of *AP* and the Planudean Anthology, it would have taken an expert a good deal of time to establish the full extent of *AP*'s superiority. Erasmus and More may not have fully appreciated it themselves.
22. His ever straitened financial circumstances make it unlikely that Erasmus could have bought it at anything approaching a fair market value.
23. Martin Lowry, *The World of Aldus Manutius: Business and Scholarship in Renaissance Venice* (Ithaca, NY, 1979), 232–33.

INDEX

This Book Was Reset on February 23, 2006
at Italica Press, New York, New York. It
Was Set in Times Roman and Was
Printed on 60-lb Natural Paper
by BookSurge
USA/EU

Made in United States
Orlando, FL
04 November 2026

90201152R00127